Chemistry

AN ELECTRONIC AND STRUCTURAL APPROACH

Chemistry
AN ELECTRONIC AND STRUCTURAL APPROACH

H G Wallace MA
Head of the Chemistry Department, St Mary's College, Crosby

V W Cowell BSc
Head of the Science Department, John Hamilton High School, Liverpool

Edited by

W E Addison BSc PhD FRIC
Institute of Science and Technology, University of Manchester

John Murray 50 Albemarle Street London

Made and printed in Great Britain by
William Clowes and Sons Limited, London and Beccles

7195 1917 9

Preface

In the first six chapters of this volume we have attempted to sketch part of the conceptual framework which is needed for an understanding of modern chemistry. We have taken a predominantly 'structural' viewpoint, the word 'structural' being interpreted in its widest sense to include electronic, crystal, molecular and nuclear structure. In subsequent chapters we have applied these structural principles to the comparative chemistry of the elements and to three aspects of the chemistry of carbon.

Our aim has been to organize the facts of descriptive chemistry into a meaningful pattern. In order to keep this volume reasonably short we have confined ourselves to a largely theoretical approach. Therefore, to be of maximum value educationally, the material in the text needs to be supplemented by an appropriate practical course and by suitable background aids such as films and filmstrips. Appropriate examples of the latter are listed at the end of each chapter. Much useful practical material will be found in the practical books of the Chemical Bond Approach (CBA) and CHEM Study projects and also in the materials, soon to be published, of the Nuffield A level Physical Science project.

In trying to achieve clarity and concision, we may sometimes appear over-dogmatic. We hope that teachers of chemistry using this book will correct this impression where necessary. They will no doubt point out to their students that the 'conceptual framework' is largely made up of theories or models and these, in the last analysis, are simply 'fruitful conjecture'.

We hope that this book will be of value to sixth form students of chemistry and physical science, to those studying for the Oxford and Cambridge Scholarship Examinations, to students in colleges of education and to first year university students.

We would like to express our gratitude to Mrs. H. Williams and Mrs. G. Edey, who typed the manuscript, and to Mr. J. G. Stark, MA, Mr. V. C. O'Donnell, BSc, and Mr. J. R. McCann, BSc, who read the proofs.

We are especially indebted to our editor, Dr. W. E. Addison, with whom we had many hours of useful discussion and who made numerous suggestions for improving the text at various stages in its production.

H.G.W.
V.W.C.

July 1969

Contents

1 The structure of matter   1

2 The Periodic Classification   17

3 The electronic theory of valency   27

4 Bonding and periodicity   55

5 The solid state   70

6 The nucleus   90

7 Hydrogen and the s-block elements   101

8 The p-block elements   117

9 The d-block elements   175

10 Isomerism   202

11 Application of valency theory to organic chemistry   224

12 Polymers   254

Index   267

Plates I to V appear between pages 72 and 73

Units

SI units are used throughout the text, e.g. energy changes are expressed in kJ mol^{-1} and atomic dimensions in nanometres, nm (1 nm $= 10^{-9}$ metres).

1. The structure of matter

The atomic theory

Whether matter is continuous or discrete is a question which has interested man for centuries. Until the dawn of experimental science proper, this interest was philosophical. It arose out of man's quest for intelligibility in his experience. John Dalton was the first to formulate an atomic theory of matter in order to explain experimental results which had been obtained at the time, and to unify in a predictive and fruitful way the chemistry of his day.

Much work had been done on the reacting masses and the composition by weight of compounds. This led to the enunciation of the four laws of chemical combination:

(1) The law of conservation of mass
(2) The law of constant composition
(3) The law of multiple proportions
(4) The law of reciprocal proportions

These laws were empirical generalizations. Dalton put forward his atomic hypothesis to explain them.

The main postulates of his theory were:

(1) All matter is composed of discrete, indivisible particles called atoms.
(2) Atoms are indestructible.
(3) Atoms of the same elements are alike in every respect, most significantly in mass.
(4) Atoms can combine together in definite numbers to form compound atoms or molecules.

The first and second postulates 'explain' the law of conservation of mass. Postulates (3) and (4) account for the law of constant composition. If it is assumed that when atoms combine they do so in small whole numbers characteristic of the elements concerned, then the laws of multiple and reciprocal proportions are readily shown to follow.

Dalton's theory, therefore, was an important step forward, leading
1*

eventually to the development of a system of chemical symbols, the concepts of equivalent weight, valency and atomic and molecular weights. However, like other scientific theories the atomic theory was an approximation and it was by the investigation of its limitations and assumptions that later progress in the understanding of structural chemistry took place.

Natural radioactivity

About one quarter of the known elements undergo spontaneous transformation; e.g. one form of rubidium gradually changes into strontium:

rubidium $\longrightarrow$ strontium + radiation
parent element daughter element

As the transformation is occurring radiation is emitted. The presence of the radiation can be shown by its effect on a photographic plate which behaves as if it had been exposed to visible light. If the radiation is allowed to fall on to a fluorescent screen (i.e. a screen covered with zinc sulphide), the screen glows. The process outlined above is called radioactive disintegration. The emitted radiation can penetrate considerable thicknesses of matter which are impervious to light; this indicates its high energy. If passed through gases it considerably increases their electrical conductivity. This latter point resulted in the electrical properties of the radiation being investigated, with the following very important results:

If a narrow beam of radiation from a radioactive source such as francium is sent between charged plates, then the beam separates into three components as shown in Fig. 1.1. β-rays are deflected towards the

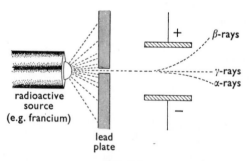

Fig. 1.1

positive plate, showing that presumably they are negatively charged. γ-rays pass through undeflected and so are uncharged. α-rays are deflected towards the negative plate and are, therefore, positively charged.

Microscopic examination of the glow emitted by a zinc sulphide screen when the radiation impinges upon it, shows myriads of tiny flashes of light. This suggests (at least as far as absorption by zinc sulphide is concerned) that the radiation is made up of streams of particles. The evidence so far considered suggests that Dalton was wrong on two points: atoms are not indestructible since apparently some of them can break down spontaneously; atoms are not structureless or 'ultimate' particles, since other particles, viz. α- and β-rays, are given off when they disintegrate.

Unit charge

Since two of the particles considered above are charged the question naturally arises—is electricity itself discrete or continuous? Millikan investigated this problem by his well-known 'oil drop' experiments. Droplets of oil are squirted between two horizontal plates. The droplets are charged by exposing them to X-rays. The rate of fall of a given drop is then observed microscopically and measured for the plates charged and uncharged. Then from the two rates, the known properties of the oil and the gas and the charge on the plates, the magnitude of the charge on each drop can be calculated. All charges prove to be multiples of a fixed minimum charge known as unit charge.

Induced radioactivity

The normally stable elements are non-radioactive because they do not possess sufficient energy to emit the highly energetic radioactive rays. However, if they are bombarded with high energy particles they become radioactive and often begin to emit α- and β-rays. Many of them also emit other particles which are not found in natural radioactivity. The most important of these are the proton and the neutron. The properties of these particles and of α- and β-rays have been investigated and may be summarized thus:

Particle	Charge (compared to unit charge)	Mass (compared to neutron)	Alternative name
Alpha	$+2$	4	Helium nucleus
Beta	-1	$\frac{1}{1840}$	Electron
Gamma	0	0	Energy
Neutron	0	1	—
Proton	$+1$	1	Hydrogen nucleus

As far as the chemist is concerned the most important particles, in terms of which the structure of the atom can be described, are the electron, the proton and the neutron.

The structure of the atom

Discharge through gases

If a gas at low pressure is subjected to an electrical discharge in a sealed tube then the following results are obtained:

(i) A beam of negatively charged particles is emitted from the cathode. These are deflected away from a negatively charged plate and can be shown to be low energy electrons.

(ii) If the anode is perforated, a beam of positive rays is emitted which travels in the opposite direction to the electron beam. This positive beam is deflected away from a positively charged plate though to a much smaller extent than the negative beam. The particles in the positive beam have masses corresponding to the atomic or molecular weight of the gas in the tube. They are positively charged atoms or molecules, i.e. gaseous ions. These gaseous ions can be produced in other ways; e.g. if a jet of gas is blown over a hot wire, ionization of the gas occurs.

The above results suggest that the atoms of the gas are made up of negatively charged electrons and much heavier positively charged particles.

Positive ray analysis

The masses of the particles present in the positive rays can be determined by a technique known as positive ray analysis. Fig. 1.2 is a schematic diagram of the apparatus used. Atoms or molecules of the gas are ionized by the hot wire as shown. Some of the positive ions move through the slits where they are accelerated by an electric field (not shown). The beam of positive particles is deflected, first by the electric field and then by the magnetic field.

Particles of high velocity are deflected to a small extent by the electric field and to a relatively large extent by the magnetic field. Particles of low velocity are deflected to a large extent by the electric field and to a relatively small extent by the magnetic field. The other variables which affect the degree of deflection are the charge and mass of the particles. By adjusting the strength of the electric and magnetic fields, therefore, all particles of the same charge to mass ratio can be focussed at the same point. Since the charge must be a multiple of unit charge, the masses of the particles can be estimated from the position at which the beam strikes the plate.

When neon is examined in the mass spectrograph two lines are obtained, one a bright line corresponding to a mass of 20 and the other a fainter line corresponding to a mass of 22. Again, chlorine gives lines corresponding to masses 37 and 35, the latter being about three times as intense as the former. This gives an average atomic mass of

$$(37 + 35 \times 3)/4 = 35 \cdot 5$$

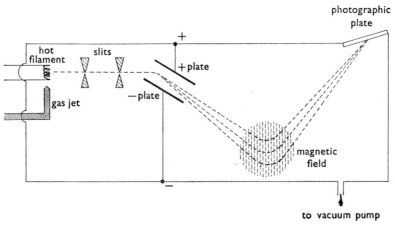

Fig. 1.2 *The mass spectrograph*

which corresponds to the chemical atomic weight of the element. About eighty per cent of the naturally occurring elements have atoms which differ in mass. These elements are said to be made up of *isotopes*, each isotope consisting of atoms of the same mass. Another of Dalton's assumptions is proved inaccurate—atoms of the same element can have different masses.

The Rutherford atom

Lord Rutherford examined a beam of α-particles which had been passed through various metal foils. The sheet of foil had a thickness corresponding to many layers of atoms. It was to be expected that the α-particles would be scattered in at least two ways: either by elastic impacts with any heavy particles present in the foil, or by repulsion by any positively charged particles present. The result was rather unexpected. A large fraction of the incident beam passed through the foil without any significant scattering, i.e. as far as the α-particles were concerned the metal was largely empty space. The few α-particles which were scattered were deflected widely from their original path. The *scattering* particles then were presumably small and therefore were seldom struck, but when struck were highly effective scattering centres, i.e. they were massive, or highly charged, or both.

From the result of his scattering experiments Rutherford was able to calculate the charge on the scattering centres in terms of unit charge; e.g. he found that platinum had a value of 78, nickel 29 and so on. Rutherford proposed the following model for atomic structure. The atom consists of a tiny central *nucleus*, positively charged, in which practically all the mass of the atom is concentrated. The nucleus is surrounded by a sufficient number of *electrons*, of almost negligible mass, to neutralize the positive charge on the nucleus. The volume of the atom depends upon the size of the electron field around the nucleus.

Atomic number

The results of mass spectroscopy indicate that atomic weight is not a defining characteristic of an element since the same element can have atoms of different masses. The *charge* on the nucleus, however, *is* characteristic of atoms of a particular element. Although chlorine occurs as two isotopes of mass 35 and 37, the nuclear charge on these is the same, namely 17. The nuclear charge is called the *atomic number* and may be defined as the number of unit positive charges on the nucleus of an atom. It will be seen later that it is also the number of surrounding orbital electrons in the neutral atom, or the ordinal number of the element in the Periodic Classification.

Structure of the nucleus

Any proposed theory of nuclear structure must explain the following observations:

(1) The nucleus is positively charged, the number of unit charges being equal to the atomic number.
(2) Nearly all the mass of the atom, apart from a small electronic mass, is concentrated in the nucleus.
(3) The relative nuclear masses are nearly whole numbers.
(4) The nucleus can undergo radioactive transformation, certain particles being ejected or added, to produce new nuclei of new mass or charge or both.
(5) The element to which a particular nucleus belongs is determined by the charge on the nucleus, not its method of formation or mass.

Structures made up of two fundamental particles, *protons* and *neutrons*, allow intepretation of these facts.

Protons carry unit positive charge and are responsible for the positive charge on the nucleus. The number of protons equals the number of positive charges, which in turn equals the atomic number of the element. Since the proton and the neutron have approximately unit mass compared to the very small relative mass of the electron, $\frac{1}{1840}$, then the nucleus accounts for nearly all the mass of the atom. Again the relative nuclear masses are approximately equal to the total number of protons and neutrons present, and so are approximately whole numbers. The number of protons in the nucleus determines the number of electrons in the electrically neutral atom. The electron field forms the 'outside' of the atom and so determines how particular atoms interact with others, i.e. it determines chemical properties. The number of electrons in the neutral atom equals the number of protons in the nucleus, which equals the atomic number and, therefore, defines atoms of a particular element.

Examples Hydrogen has atomic number 1, mass number 1. The nucleus is, therefore, made up of one proton, charge $+1$, mass 1.

Helium has atomic number 2, mass number 4. The nucleus consists of two protons, charge $+2$, mass 2, and two neutrons, charge 0, mass 2.

Sodium has atomic number 11, mass number 23. The nucleus, therefore, has eleven protons, charge $+11$, mass 11, and twelve neutrons, charge 0, mass 12.

In general, if mass number $=A$ and atomic number $=Z$, then

$$n = A - Z$$

where n is the number of neutrons.

Isotopes

Chlorine has two isotopes of mass 35 and 37. They both have the same atomic number, 17. (All isotopes must have the *same atomic number* since they are atoms of the *same element*.) The first isotope, therefore, contains 17 protons and 18 neutrons in its nucleus, giving a total charge of 17 and a total mass of $18+17=35$. The second isotope has 17 protons and 20 neutrons in its nucleus, giving the same charge as the first, 17, but a mass of $20+17=37$. Isotopes are usually denoted by putting the mass number at the top left of the symbol for the element and the atomic number at the bottom left, thus:

$$\text{Chlorine mass 35} \quad {}^{35}_{17}\text{Cl}$$
$$\text{Chlorine mass 37} \quad {}^{37}_{17}\text{Cl}$$

The chemical atomic weight of an element, as shown earlier in this chapter, is equal to the average of the isotopic masses present in a naturally occurring sample of the element. Thus:

$${}^{35}_{17}\text{Cl} \quad \text{relative abundance 75\% approx.}$$
$${}^{37}_{17}\text{Cl} \quad \text{relative abundance 25\% approx.}$$

$$\text{Atomic weight} = \left(\frac{75}{100} \times 35\right) + \left(\frac{25}{100} \times 37\right) = 35 \cdot 5$$

Nuclear transformation

This occurs by the emission or absorption of particles by the nucleus and can be represented by equations similar to chemical equations, if the notation of the last section is used; for example,

$$^{238}_{92}U \longrightarrow {}^{4}_{2}He + {}^{234}_{90}Th + energy$$

Both mass and charges balance. The equation means that uranium 238 disintegrates giving a helium nucleus, i.e. an α-particle, the daughter element thorium, and energy (γ-rays). Chapter 6 gives a more detailed discussion of nuclear transformations.

The structure of the electron field

Rutherford thought that the electrons were revolving around the central nucleus like planets around a central sun. The force of attraction between electrons and nucleus provides the centripetal acceleration. However, according to classical theory, such an accelerating electron would continually give out energy as electromagnetic radiation. Consequently the electron would continuously lose energy and so would spiral into the nucleus. The Rutherford atom was inherently unstable, which does not agree with the observed properties of most atoms.

The Bohr atom

In order to explain such phenomena as the photo-electric effect and the spectral distribution of black-body radiation at different temperatures, Max Planck put forward the hypothesis that energy was emitted and absorbed in discrete packets called *quanta*, rather than continuously. The size of a quantum was not fixed but depended upon the frequency of the radiation, according to the equation

$$E = h\nu$$

where E = the energy corresponding to a quantum
   h = a constant known as *Planck's constant*
   ν = the frequency of the radiation.

If the spectrum from an incandescent gaseous element is examined, it is found not to be continuous but made up of a number of lines. It is called a *line spectrum*. This line spectrum cannot be explained using the Rutherford atom.

To explain these atomic line spectra Niels Bohr developed an atomic model based upon the quantum theory. He put forward two main postulates:

(1) The electrons can revolve around the nucleus only in certain definite orbits corresponding to certain 'allowed' energy states.

(2) Only those orbits could occur for which the angular momentum was an integral multiple of $h/2\pi$; that is,

$$\text{angular momentum} = n\frac{h}{2\pi}$$

where n is an integer known as a *quantum number*.

An analogy may make this clearer. Consider a valley with terraced sides, as shown in Fig. 1.3. The base of the valley represents the nucleus

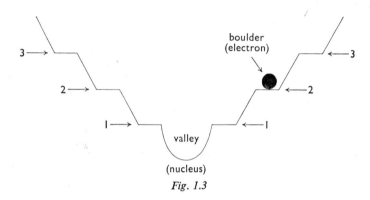

Fig. 1.3

and the boulder represents an electron. If the boulder moves towards the valley base it must move from level 2 to level 1. It cannot remain between levels. On moving from level 2 to level 1 the boulder loses a definite amount of potential energy corresponding to the difference in vertical height between the two levels. In an incandescent gas the electrons are excited from the ground state to higher energy levels. If the electron then moves back towards the nucleus it gives out energy as electromagnetic radiation. However, since the electron like the boulder can exist only in certain fixed energy levels, when it moves towards the ground state it must do so in a series of discrete energy jumps analogous to the potential energy jumps of the boulder. However, since $E=h\nu$, then for a definite ΔE_1,

$$\Delta E_1 = h\nu_1$$

and hence

$$\nu_1 = \frac{\Delta E_1}{h}$$

Thus for each definite energy jump towards the nucleus, radiation of a definite frequency will be emitted, producing the observed atomic line spectra.

Quantum numbers

When the spectrum of hydrogen is examined, groups of spectral lines are found corresponding to electron jumps between the main energy levels. These main energy levels are denoted by the *principal quantum number n*, which has values 1, 2, 3, etc. Examination of the spectrum with a spectroscope of high resolving power shows further splitting of the lines corresponding to main energy levels. This fine structure results from the division of the main energy levels into sub-levels denoted by the *sub-level quantum number l*. When the principal quantum number has the value n then l can have values

$$n-1, n-2, \ldots, 0$$

In the presence of a magnetic or electric field further splitting of the sub-level spectral lines occurs; this is known as the *Zeeman effect*. The *magnetic quantum number m* is used to denote the resulting levels. If the sub-level quantum number has the value l, then the magnetic quantum number can have the values

$$l, \ldots, 0, \ldots, -l$$

Finally examination of single lines in the sub-level series in the spectra of elements such as the alkali metals shows further splitting into doublets or pairs of lines. This is denoted by the spin quantum number s which may take the values $\frac{1}{2}$ and $-\frac{1}{2}$ for each value of m. A physical interpretation of quantum numbers and examples of their application are given later.

De Broglie's equation

Bohr's atomic model achieved considerable success in interpreting experimental results. Theoretically, however, it seemed rather artificial in that Bohr's assumptions were introduced as a modification of the classical picture, with little apparent theoretical justification. De Broglie's work was an attempt to remedy this. He suggested that all particles had an associated wave character. Light, for example, was best considered as a wave *during propagation*, but as particulate (made up of quanta or photons) *during emission and absorption*. The behaviour of light is described by two fundamental equations:

(1) Einstein's equation:

$$E = mc^2$$

where E=energy, m=mass of particle, c=velocity.
(2) Planck's equation:

$$E=h\nu$$

where E=energy, h=Planck's constant, ν=frequency.

Since $c = \nu\lambda$

where $\lambda =$ the wavelength, it follows that

$$E = mc^2 = h\frac{c}{\lambda}$$

Hence

$$\lambda = \frac{h}{mc}$$

De Broglie suggested that this equation could be applied to any particle; i.e. in general,

$$\lambda = \frac{h}{mv}$$

where $\lambda =$ wavelength of associated wave

$h =$ Planck's constant

$m =$ mass of particle

$v =$ velocity of particle.

The wave character of free-moving electrons has been demonstrated experimentally; an electron beam can be made to undergo diffraction under the appropriate conditions.

If an electron is moving around the nucleus of an atom then the associated wave will be a standing wave. This means that allowed orbits will be those having a circumference which can contain an exactly whole number of wavelengths, i.e. for which the following condition holds:

$$2\pi r = n\lambda$$

where $r =$ the radius of the orbit

$\lambda =$ the wavelength of the standing wave

$n =$ an integer.

If this condition is not fulfilled the electron wave will destroy itself by interference. An example of an 'allowed' and a 'not-allowed' orbit is shown in Fig. 1.4.

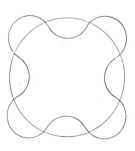
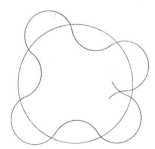

allowed' orbit   'not allowed' orbit

Fig. 1.4

The introduction of the principal quantum number n now appears quite reasonable and is a direct consequence of the wave character of the electron.

Wave-mechanical model of electron distribution

A wave motion can be described mathematically by means of a wave function Ψ (psi). This describes the way in which the amplitude of the wave varies with its spatial coordinates and the kinetic and potential energy of the wave. The intensity of a beam of light (number of photons per unit volume) is proportional to Ψ^2 where Ψ represents the amplitude of the electrical vector. Using the de Broglie equation $\lambda = h/mv$, Schrödinger derived a wave equation for an electron associated with the hydrogen nucleus. Just as Ψ^2 for light is proportional to the 'probability density' of photons, then Ψ^2 in the Schrödinger equation is proportional to the 'probability density' for electron 'matter' waves. The probability of finding an electron in a region of space $dx\,dy\,dz$ is proportional to $\Psi^2\,dx\,dy\,dz$. In other words over a period of time Ψ^2 indicates the distribution of electrical charge about the nucleus. In more concrete terms, if the electron is regarded as a negative charge cloud then Ψ determines the shape of the cloud and $\Psi^2\,dx\,dy\,dz$ the charge density.

Solution of the wave equation, therefore, gives the 'probability distribution' for the electron. Each solution of the equation is characterized by four quantum numbers. These may be considered to have the following significance:

(i) *The principal quantum number n* indicates the main energy level to which the electrons belong.

(ii) *The sub-level quantum number l* indicates the sub-level to which the electron belongs. Each value of this is associated with a 'probability distribution' of a particular shape.

(iii) *The magnetic quantum number m* is related to the orientation of 'probability distributions' of electrons with a given sub-level number l.

(iv) *The spin quantum number s* indicates the direction of electron spin.

Consider the hydrogen atom in its ground state. Then the orbital electron is in the first main energy level, that is

$$n = 1$$

Now l can have values $n-1, \ldots, 0$. Therefore in this case

$$l = 0$$

Again, m can have values $l, \ldots, 0, \ldots, -l$. Therefore in this case

$$m = 0$$

The quantum number s can be either $+\frac{1}{2}$ or $-\frac{1}{2}$. Since there is only one electron let $s = +\frac{1}{2}$. The electron is now labelled by the four quantum numbers

$$n = 1, \quad l = 0, \quad m = 0, \quad s = +\frac{1}{2}$$

According to the Bohr model this electron would be orbiting the nucleus in a definite path of lowest energy. According to the wave mechanical model only the probability of finding the electron at a particular point in space can be calculated. In Fig. 1.5 the probability is indicated by

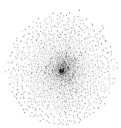

Fig. 1.5

dots so that the density of the dots increases with increasing probability.

Naturally, since the electron is negatively charged and the nucleus positively charged the probability is a maximum at the nucleus and decreases moving out radially from the centre. Thus a graph of distance r from nucleus against the probability of the electron being at that particular point, is of the form shown in Fig. 1.6.

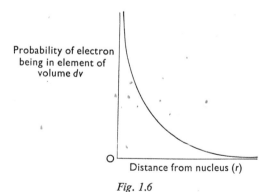

Probability of electron being in element of volume dv

Distance from nucleus (r)

Fig. 1.6

However, if the probability of locating the electron in a *spherical shell* distance r from the nucleus is plotted against r then a maximum is obtained, since, although the probability decreases with distance from nucleus, the volume of the shell increases as the square of the distance.

The volume of a spherical shell of thickness dr is given by $4\pi r^2\, dr$. Fig. 1.7 shows the resulting graph.

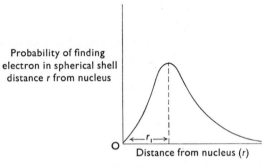

Probability of finding
electron in spherical shell
distance r from nucleus

r_1

Distance from nucleus (r)

Fig. 1.7

Not surprisingly the distance r_1 at which this maximum occurs corresponds to the radius of the electron orbit in the Bohr model. Then, ignoring the region of low probability, the probability distribution shown in Fig. 1.5 can be redrawn as in Fig. 1.8. There is a high probability of

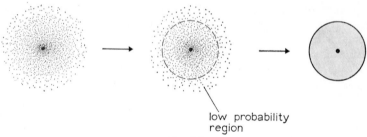

low probability
region

Fig. 1.8

finding the electron in the shaded volume. This means that the electron can be considered as a cloud of negative electricity occupying a definite (though arbitrarily defined) region of space relative to the nucleus. The spherical probability distribution or charge cloud discussed above applies only to electrons with sub-level quantum number $l=0$. Other values of l give charge clouds of different shapes. Instead of using values of l equal to 0, 1, 2 and 3, it is usual for chemists to use the spectroscopic notation s, p, d and f. These are the initial letters of the words *sharp, principal, diffuse* and *fundamental*—terms used to describe the spectral lines. Thus an electron with $n=1$, $l=0$ would be called a 1s electron. If $n=3$, $l=2$ then this would be a 3d electron. The region of space which can be (but not necessarily is) occupied by an electron is called an *orbital.** Therefore, one could also refer to the 1s orbital, the 3d orbital, etc.

Shapes and orientations of the orbitals

(1) *The s orbitals*   These were discussed in the last section. An s orbital is spherically symmetrical about the nucleus. Because an s orbital possesses this symmetry there is no splitting into orbitals of different energy in a magnetic field.

(2) *The p orbitals*   These are dumb-bell shaped and there are three of them, each one lying along a Cartesian axis, as shown in Fig. 1.9.

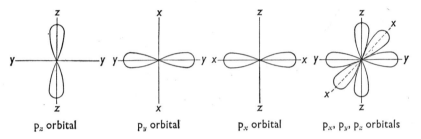

p_z orbital p_y orbital p_x orbital p_x, p_y, p_z orbitals

Fig. 1.9

In the absence of a magnetic or electric field the p orbitals all have the same energy. The p level is said to show *three-fold degeneracy*: this means that there are three solutions to the wave equation for the same energy value. In the presence of a magnetic or electric field the three orbitals are affected differently because of their different orientation. Each of the orbitals has different energy and this results in a splitting of the spectral line to give lines corresponding to the different energies. The threefold degeneracy is said to be *resolved*.

(3) *The d orbitals*   The d orbitals are four-lobed, apart from the d_{z^2}, and have the orientations shown in Fig. 1.10. The d level is five-fold degenerate in the absence of magnetic or electric fields.

* The wave function Ψ is the product of two terms: (i) the *angular function*, which gives information about how the probability of finding an electron in an element of volume dv varies with direction from the nucleus; and (ii) the *radial function*, which gives information about how the probability varies with distance from the nucleus. The term 'orbital' should, strictly speaking, refer to a particular solution of the wave equation, and this is the product of these two terms. As such an 'orbital' cannot be pictured as a physical reality, so throughout the text orbital will also be used to refer to the associated 'charge cloud'. This, of course, is a representation of the $\Psi^2 dv$ function as explained above.

(4) *The f orbitals*   The shape and orientation of these orbitals is beyond the scope of this book. It is sufficient to note that there are seven f orbitals of equal energy. The f level is seven-fold degenerate.

Each orbital in all the above cases can contain a maximum of two electrons of opposing spin.

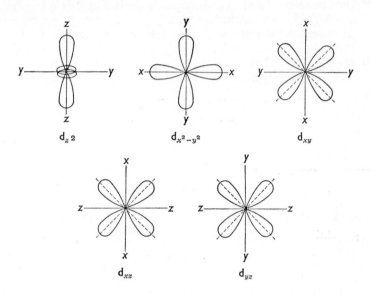

Fig. 1.10

References

Books

General Chemistry, Steiner and Campbell (Macmillan)
General and Inorganic Chemistry, P. J. Durrant (Longmans)
Modern Valency Theory, G. I. Brown (Longmans)
Physical Chemistry, W. J. Moore (Longmans)
Valency and Molecular Structure, Cartmell and Fowles (Butterworths)
Electronic Structure, Properties and the Periodic Law, H. H. Sisler (Chapman and Hall)

Films

The Hydrogen Atom As Viewed By Quantum Mechanics (Standard version), G. C. Pimentel—a CHEM Study Film. Available from Sound-Services Ltd., cat. no. 4148/999

2. The Periodic Classification

From the beginning of the nineteenth century many attempts to correlate the properties of the elements with their atomic weights were made. Most of these were unsuccessful, though in 1864 John Newlands listed the elements in order of their atomic weights and showed that every eighth element appeared to fall into a family possessing similar properties. He gave this phenomenon the rather fanciful name the 'Law of Octaves', and his work received little attention. In 1870 Dmitri Mendeleev formulated his periodic law which stated that the chemical and physical properties of the elements are periodic functions of their atomic weights. He was able to draw up a periodic table which classified the elements into groups or sets with similar properties.

It has been pointed out, however, that atomic weight is not the fundamental characterizing property of an element, and so there are various anomalies in Mendeleev's classification, though this does not detract from the important systematizing effect the classification had on inorganic chemistry.

The modern Periodic Table is based upon current theories of atomic structure and the periodic law can be reformulated in terms of the more fundamental characteristic of elements, atomic number: the chemical and physical properties of the elements are periodic functions of their atomic numbers. The atomic number determines the number of orbital electrons and hence, as will be explained below, the structure of the electron field. It is the electron field of an atom which interacts with that of other atoms, so that atomic number and properties will be closely related.

The Pauli exclusion principle (1925)

This simply states that no two electrons in the same atom can have the same value for each of the four quantum numbers. Using this and assuming the relationship between the quantum numbers stated in Chapter 1, the maximum number of electrons contained in a main energy level or a sub-level can be deduced as shown in Table 2.1. This

n	$l, (n-1, n-2, \ldots, 0)$	$m, (-l, \ldots, 0, \ldots, +l)$	s	Number of electrons
1	0 (s)	0	$\frac{1}{2}, -\frac{1}{2}$	Two 1s electrons TOTAL 2
2	0 (s)	0	$\frac{1}{2}, -\frac{1}{2}$	Two 2s electrons
	1 (p)	1	$\frac{1}{2}, -\frac{1}{2}$	} Six 2p
		0	$\frac{1}{2}, -\frac{1}{2}$	electrons
		−1	$\frac{1}{2}, -\frac{1}{2}$	TOTAL 8
3	0 (s)	0	$\frac{1}{2}, -\frac{1}{2}$	Two 3s electrons
	1 (p)	1	$\frac{1}{2}, -\frac{1}{2}$	} Six 3p
		0	$\frac{1}{2}, -\frac{1}{2}$	electrons
		−1	$\frac{1}{2}, -\frac{1}{2}$	
	2 (d)	2	$\frac{1}{2}, -\frac{1}{2}$	}
		1	$\frac{1}{2}, -\frac{1}{2}$	
		0	$\frac{1}{2}, -\frac{1}{2}$	Ten 3d electrons
		−1	$\frac{1}{2}, -\frac{1}{2}$	
		−2	$\frac{1}{2}, -\frac{1}{2}$	TOTAL 18
4	0 (s)	0	$\frac{1}{2}, -\frac{1}{2}$	Two 4s electrons
	1 (p)	1	$\frac{1}{2}, -\frac{1}{2}$	} Six 4p
		0	$\frac{1}{2}, -\frac{1}{2}$	electrons
		−1	$\frac{1}{2}, -\frac{1}{2}$	
	2 (d)	2	$\frac{1}{2}, -\frac{1}{2}$	}
		1	$\frac{1}{2}, -\frac{1}{2}$	
		0	$\frac{1}{2}, -\frac{1}{2}$	Ten 4d electrons
		−1	$\frac{1}{2}, -\frac{1}{2}$	
		−2	$\frac{1}{2}, -\frac{1}{2}$	
	3 (f)	3	$\frac{1}{2}, -\frac{1}{2}$	}
		2	$\frac{1}{2}, -\frac{1}{2}$	
		1	$\frac{1}{2}, -\frac{1}{2}$	
		0	$\frac{1}{2}, -\frac{1}{2}$	Fourteen 4f electrons
		−1	$\frac{1}{2}, -\frac{1}{2}$	
		−2	$\frac{1}{2}, -\frac{1}{2}$	
		−3	$\frac{1}{2}, -\frac{1}{2}$	TOTAL 32

Table 2.1

shows the quantum numbers for electrons in the first four main energy levels. The total number of electrons for each level forms the following series:

$$2 \qquad 8 \qquad 18 \qquad 32$$

i.e. $\qquad 2 \times 1^2 \quad 2 \times 2^2 \quad 2 \times 3^2 \quad 2 \times 4^2$

It might be expected that level 5 could contain $2 \times 5^2 = 50$ electrons. However, as the atomic number of the element increases beyond a certain point, the element becomes less stable (i.e. becomes radioactive) so that there are only a hundred or so known elements, those with highest atomic numbers being man-made. For these elements the fifth main energy level does not develop beyond the f sub-level, the sixth does not develop beyond the d sub-level and the seventh does not develop beyond the s sub-level. For the sub-levels, irrespective of main quantum number, the maximum number of electrons that can be present is given by the series

	s	p	d	f
	2	6	10	14
i.e.	2×1	2×3	2×5	2×7

The 'Aufbau' principle

This is the building-up principle. For each proton added to the nucleus the atomic number increases by unity and so one passes from one element to the next. For an electrically neutral atom, addition of a proton to the nucleus must be accompanied by the addition of an electron to the electron field. The maximum number of electrons which each sub-level and main level can accommodate is known from the previous discussion; therefore, if the relative magnitudes of the energy levels were known it should be possible to build up the electronic structures of all the elements. Fortunately this order of magnitude falls into a regular pattern as shown in Fig. 2.1. This gives the simplest form for writing

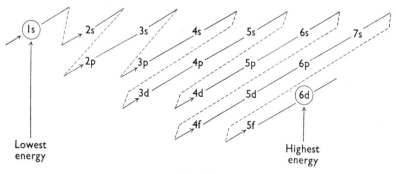

Fig. 2.1

down and hence remembering this energy order; it is easier to use, perhaps, if redrawn as in Fig. 2.2.

Now assuming that the electrons will enter the lowest possible energy levels, except when the atom is in an excited state (e.g. in an incandescent gas), the electronic structures for atoms of the elements in the ground state can be built up as follows.

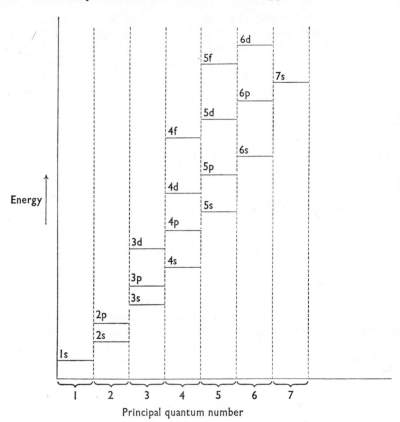

Fig. 2.2 (The energy scale gives only the order of energies; the magnitude of the energy is not indicated.)

Hydrogen has atomic number 1, and an electronic structure of one electron in the 1s level. If a superscript is used to denote the number of electrons and the chemical symbol is used to denote the nucleus then the structure of hydrogen is

$$H\ 1s^1$$

Helium has atomic number 2 and so possesses two electrons both in the lowest energy level. Its structure is

$$He\ 1s^2$$

This process can be continued, an electron being fed into the lowest available energy level as the atomic number increases by unity. For the first three periods of the Periodic Table this would give:

1. H $1s^1$
2. He $1s^2$ } Period 1

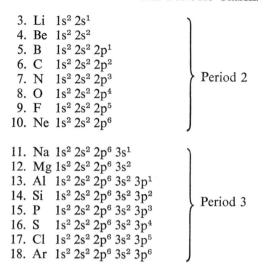

3. Li $1s^2\,2s^1$
4. Be $1s^2\,2s^2$
5. B $1s^2\,2s^2\,2p^1$
6. C $1s^2\,2s^2\,2p^2$
7. N $1s^2\,2s^2\,2p^3$
8. O $1s^2\,2s^2\,2p^4$ } Period 2
9. F $1s^2\,2s^2\,2p^5$
10. Ne $1s^2\,2s^2\,2p^6$

11. Na $1s^2\,2s^2\,2p^6\,3s^1$
12. Mg $1s^2\,2s^2\,2p^6\,3s^2$
13. Al $1s^2\,2s^2\,2p^6\,3s^2\,3p^1$
14. Si $1s^2\,2s^2\,2p^6\,3s^2\,3p^2$
15. P $1s^2\,2s^2\,2p^6\,3s^2\,3p^3$ } Period 3
16. S $1s^2\,2s^2\,2p^6\,3s^2\,3p^4$
17. Cl $1s^2\,2s^2\,2p^6\,3s^2\,3p^5$
18. Ar $1s^2\,2s^2\,2p^6\,3s^2\,3p^6$

If the periods are arranged in horizontal rows, this gives a modern form of the Periodic Table as shown in Table 2.2. In this form of the table the notation IM, IIM, IIIM and so on is used for the main groups, i.e. the s- and p-block elements. The elements produced by the development of a d sub-level are denoted by groups IT_d, IIT_d etc. and those produced by the development of an f sub-level by groups IT_f, IIT_f etc. This notation is preferred by the authors for the following reasons:

(1) The group number for the s- and p-block elements corresponds to the number of electrons in the outer main energy level. M indicates that a main energy level is being developed.

(2) For the d- and f-block elements the group number corresponds to the number of electrons in the sub-level being developed, T indicates an inner sub-level and the suffix (d or f) indicates the type of sub-level.

(3) For the d- and f-block elements the notation emphasizes that the important relationships are vertical (within a group) and horizontal (across a transition element series).

In more conventional forms of the table the notation is IA, IIA, IIIA, IVA, etc. for the s- and p-block elements and for the d-block elements (starting with the scandium group) IIIB, IVB, VB, VIB, VIIB; then the iron, cobalt and nickel groups are grouped together as Group VIII, followed by the copper and zinc groups which are labelled IB and IIB. The advantage of this transition series notation is that it is supposed to indicate cross-relationships, e.g. between Group IA (the alkali metals) and Group IB (the copper group). These relationships are often unimportant, however, when compared with the vertical and horizontal relationships within the d-block elements. Again the more conventional

Periodic table of the elements (electron configurations).

Group labels across the top: **I, II, III, IV, V, VI, VII**

$$
\begin{array}{cc}
\text{1} & \text{2}\\
\text{H} & \text{He}\\
1s^1 & 1s^2
\end{array}
$$

s-block elements

IM	IIM
3 Li — $2s^1$	4 Be — $2s^2$
11 Na — $3s^1$	12 Mg — $3s^2$
19 K — $4s^2$	20 Ca — $4s^2$
37 Rb — $5s^1$	38 Sr — $5s^2$
55 Cs — $6s^1$	56 Ba — $6s^2$
87 Fr — $7s^1$	88 Ra — $7s^2$

d-block elements

IIIT$_d$	IVT$_d$	VT$_d$	VIT$_d$	VIIT$_d$	VIIIT$_d$	VIIIT$_d$	VIIIT$_d$	IT$_d$	XT$_d$
21 Sc $3d^1 4s^2$	22 Ti $3d^2 4s^2$	23 V $3d^3 4s^2$	24 Cr $3d^5 4s^1$	25 Mn $3d^5 4s^2$	26 Fe $3d^6 4s^2$	27 Co $3d^7 4s^2$	28 Ni $3d^8 4s^2$	29 Cu $3d^{10} 4s^1$	30 Zn $3d^{10} 4s^2$
39 Y $4d^1 5s^2$	40 Zr $4d^2 5s^2$	41 Nb $4d^4 5s^1$	42 Mo $4d^5 5s^1$	43 Tc $4d^6 5s^1$	44 Ru $4d^7 5s^1$	45 Rh $4d^8 5s^1$	46 Pd $4d^{10} 5s^0$	47 Ag $4d^{10} 5s^1$	48 Cd $4d^{10} 5s^2$
57 La* $5d^1 6s^2$	72 Hf $4f^{14} 5d^2$ $6s^2$	73 Ta $4f^{14} 5d^3$ $6s^2$	74 W $4f^{14} 5d^4$ $6s^2$	75 Re $4f^{14} 5d^5$ $6s^2$	76 Os $4f^{14} 5d^6$ $6s^2$	77 Ir $4f^{14} 5d^7$ $6s^2$	78 Pt $4f^{14} 5d^9$ $6s^1$	79 Au $4f^{14} 5d^{10}$ $6s^1$	80 Hg $4f^{14} 5d^{10}$ $6s^2$
89 *Ac $6d^1 7s^2$									

p-block elements

IIIM	IVM	VM	VIM	VIIM	VIIIM
5 B $2s^2 2p^1$	6 C $2s^2 2p^2$	7 N $2s^2 2p^3$	8 O $2s^2 2p^4$	9 F $2s^2 2p^5$	10 Ne $2s^2 2p^6$
13 Al $3s^2 3p^1$	14 Si $3s^2 3p^2$	15 P $3s^2 3p^3$	16 S $3s^2 3p^4$	17 Cl $3s^2 3p^5$	18 Ar $3s^2 3p^6$
31 Ga $3d^{10} 4s^2$ $4p^1$	32 Ge $3d^{10} 4s^2$ $4p^2$	33 As $3d^{10} 4s^2$ $4p^3$	34 Se $3d^{10} 4s^2$ $4p^4$	35 Br $3d^{10} 4s^2$ $4p^5$	36 Kr $3d^{10} 4s^2$ $4p^6$
49 In $4d^{10} 5s^2$ $5p^1$	50 Sn $4d^{10} 5s^2$ $5p^2$	51 Sb $4d^{10} 5s^2$ $5p^3$	52 Te $4d^{10} 5s^2$ $5p^4$	53 I $4d^{10} 5s^2$ $5p^5$	54 Xe $4d^{10} 5s^2$ $5p^6$
81 Tl $4f^{14} 5d^{10}$ $6s^2 6p^1$	82 Pb $4f^{14} 5d^{10}$ $6s^2 6p^2$	83 Bi $4f^{14} 5d^{10}$ $6s^2 6p^3$	84 Po $4f^{14} 5d^{10}$ $6s^2 6p^4$	85 At $4f^{14} 5d^{10}$ $6s^2 6p^5$	86 Rn $4f^{14} 5d^{10}$ $6s^2 6p^6$

f-block elements

IT$_f$	IIT$_f$	IIIT$_f$	IVT$_f$	VT$_f$	VIT$_f$	VIIT$_f$	VIIIT$_f$	IXT$_f$	XT$_f$	XIT$_f$	XIIT$_f$	XIIIT$_f$	XIVT$_f$
58 Ce $4f^2 5d^0$	59 Pr $4f^3 5d^0$	60 Nd $4f^4 5d^0$	61 Pm $4f^5 5d^0$	62 Sm $4f^6 5d^0$	63 Eu $4f^7 5d^0$	64 Gd $4f^7 5d^0$	65 Tb $4f^9 5d^0$	66 Dy $4f^{10} 5d^0$	67 Ho $4f^{11} 5d^0$	68 Er $4f^{12} 5d^0$	69 Tm $4f^{13} 5d^0$	70 Yb $4f^{14} 5d^0$	71 Lu $4f^{14} 5d^1$
90 Th $5f^0 6d^2$	91 Pa $5f^2 6d^1$	92 U $5f^3 6d^1$	93 Np $5f^4 6d^1$	94 Pu $5f^6 6d^0$	95 Am $5f^7 6d^0$	96 Cm $5f^7 6d^0$	97 Bk $5f^9 6d^0$	98 Cf $5f^{10} 6d^0$	99 Es $5f^{11} 6d^0$	100 Fm $5f^{12} 6d^0$	101 Md $5f^{13} 6d^0$	102 No $5f^{14} 6d^0$	103 Lw $5f^{14} 6d^1$

Filled shells (listed by period)

- $1s^2$
- $1s^2$; $2s^2 2p^6$
- $1s^2$; $2s^2 2p^6$; $3s^2 3p^6$
- $1s^2$; $2s^2 2p^6$; $3s^2 3p^6\ 3d^{10}$
- $1s^2$; $2s^2 2p^6$; $3s^2 3p^6\ 3d^{10}$; $4s^2 4p^6\ 4d^{10}$
- $1s^2$; $2s^2 2p^6$; $3s^2 3p^6\ 3d^{10}$; $4s^2 4p^6\ 4d^{10}\ 4f^{14}$; $5s^2 5p^6\ 5d^{10}$; $6s^2 6p^6$

notation is not so directly related to the electronic structure of the elements.

Period 4 starts with the development of the 4s level to give potassium and calcium, but the next lowest energy level according to Fig. 2.2 is the 3d. So unlike Periods 2 and 3, the next ten elements (scandium to zinc) are given by the development of the penultimate 3d level. The next lowest level is the 4p which is developed to give a further six elements (gallium to krypton). Period 5 is produced in a similar manner to Period 4, by the development of the 5s, 4d and 5p levels. In Period 6, after the development of the 6s level to give caesium and barium, the remaining elements are produced by the development of the 4f, 5d and 6p levels in this order. Development of the 4f level gives an additional 14 elements in this period. The elements of Period 7 are all radioactive, as the atomic number has increased to the extent that spontaneous disintegration occurs. The structures of those elements which do exist in this period are given by the development of the 7s and 5f levels.

General characteristics of the classification

Empirical characteristics

(1) The elements in columns or *groups* show similar, though graded, chemical and physical properties. Each group forms a family of elements.

(2) Elements in a given group often have the same range of valencies or oxidation numbers.

(3) In each horizontal row or period there is a gradual change in properties from metallic to non-metallic.

(4) Periods 4 and 5 each contain 10 more elements than Periods 2 and 3. These elements show a greater similarity to each other in chemical and physical properties than the other elements of the period. The elements are called *transition* elements.

(5) Period 6 also has a transition series of 10 elements. It contains in addition a series of 14 elements which are even more similar to each other in properties. These elements are called the *lanthanides* or *rare earths*. They are said to form an inner transition series.

(6) Group 8 contains the noble gases, which are relatively unreactive and stable.

Electronic characteristics

(1) Each period is produced by the development of a main energy level. The period number is the same as the principle quantum number of the energy level being developed initially in Groups I and II.

(2) Leaving out the transition series the number of electrons in the outer shell is the same for each group, and equals the group number and

the maximum oxidation number or valency of the elements in the group. The outer shell is called the *valency shell* and the electrons in it are called the *valency electrons*.

(3) The transition series are formed by the development of a d sub-level in the penultimate shell, for example, in Period 4, shell 4 is being developed, but the transition series scandium to zinc is formed by the development of the 3d level. An inner transition series is formed by the development of an ante-penultimate f sub-level; for example, in Period 6, shell 6 is being developed, but the lanthanides are formed by the development of the 4f sub-level.

(4) The noble gases in Group VIIIM all contain eight electrons in their valency shells.

(5) The ordinal number of an element in the table is equal to the atomic number, which equals the number of protons in the nucleus.

A comparison of the electronic and empirical characteristics suggests certain obvious correlations. All elements in the same group have similar properties: all elements in the same group have valency shells with the same electronic structure. The maximum oxidation number (except in the transition series) is often characteristic of each group and equals the number of electrons in the outermost shell. The transition elements have *similar* properties and the *same* number of electrons (with a few exceptions) in the outermost shell. The lanthanides, with *very similar* properties, have identical or almost identical structures for the *two* outermost shells. Much of inorganic chemistry is aimed at discovering and elucidating these correlations. This will be attempted in more detail in Chapters 7–9.

The distribution of electrons between orbitals

The distribution of electrons in the main energy levels and sub-levels has been discussed. Their distribution between orbitals of the same energy in the same sub-level will now be considered. The structures of boron, carbon, nitrogen, oxygen, fluorine and neon are as follows:

	Atomic number	*Structure*
B	5	B $1s^2\,2p^1$
C	6	C $1s^2\,2p^2$
N	7	N $1s^2\,2p^3$
O	8	O $1s^2\,2p^4$
F	9	F $1s^2\,2p^5$
Ne	10	Ne $1s^2\,2p^6$

What is the distribution of electrons between the three degenerate 2p orbitals, $2p_x$, $2p_y$, $2p_z$? For boron there is no problem since the single electron must occupy one of the available 2p orbitals, say the $2p_x$, giving the structure

$$\text{B } 1s^2 \, 2p_x^1 \, 2p_y^0 \, 2p_z^0$$

Similarly, since each p orbital can contain a maximum of two electrons, the structures of fluorine and neon are obvious, viz.:

$$\text{F} \quad 1s^2 \, 2p_x^2 \, 2p_y^2 \, 2p_z^1$$
$$\text{Ne } 1s^2 \, 2p_x^2 \, 2p_y^2 \, 2p_z^2$$

To obtain the structures of carbon, nitrogen and oxygen the *principle of maximum multiplicity* must be used. This states that when degenerate orbitals are being filled one electron goes into each orbital so that each orbital is half full, and then the orbitals are completed. Electrons in the half-filled orbitals have the same or parallel spins. If the symbol ↑ is used to represent an electron with a certain spin and ↓ an electron with opposite spin, then the structures of B, C, N, O, F and Ne are as follows:

B	$1s^2$	↑		
		$2p_x^1$	$2p_y$	$2p_z$
C	$1s^2$	↑	↑	
		$2p_x^1$	$2p_y^1$	$2p_z$
N	$1s^2$	↑	↑	↑
		$2p_x^1$	$2p_y^1$	$2p_z^1$
O	$1s^2$	↑↓	↑	↑
		$2p_x^2$	$2p_y^1$	$2p_z^1$
F	$1s^2$	↑↓	↑↓	↑
		$2p_x^2$	$2p_y^2$	$2p_z^1$
Ne	$1s^2$	↑↓	↑↓	↑↓
		$2p_x^2$	$2p_y^2$	$2p_z^2$

Since electrons are negatively charged they will exert a repulsive force upon each other; therefore they tend to arrange themselves in space to be as far apart as possible. Thus the half-filling of the orbitals before electron pairing occurs is understandable.

The principle of maximum multiplicity can be used in a similar manner to predict the distribution of electrons in d and f orbitals, for atoms in the ground state.

References

Books

Electronic Structure, Properties and the Periodic Law, H. H. Sisler (Chapman and Hall)

2+

Structural Principles in Inorganic Compounds, W. E. Addison (Longmans)

Papers

'Nomenclature of the Periodic Table', S. S. Chissick and J. Baldwin, *Education in Chemistry*, **2**, No. 4, July 1965, pp. 181–184

Films

Chemical Families, J. L. Hallenberg and J. A. Campbell—a CHEM Study Film. Available from Sound-Services Ltd., cat. no. 4112/999

3. The electronic theory of valency

Introduction

To account for compound formation Dalton postulated the formation of compound atoms or molecules. In these, atoms were joined together in definite numbers. This implies the formation of 'bonds' of some kind between atoms. Dalton, of course, gave no explanation of the nature of these bonds but suggested that bond formation was an inherent characteristic of atoms. The development of a more detailed and sophisticated theory of atomic structure has produced the conceptual tools required for a much fuller understanding of bonding or valency.

It is usual to distinguish, from a logical point of view, certain definite types of chemical bond. This procedure will be followed in this chapter. It is important to remember, however, that the bonding in the majority of chemical compounds does not fall neatly into one of these divisions, and is more adequately described as lying somewhere between the extreme types explained below.

The ionic bond (electrovalency)

If a piece of metallic sodium is lowered into a gas-jar of chlorine, a vigorous reaction occurs and a white solid, sodium chloride, is formed. The bonding between sodium and chlorine in sodium chloride is almost purely ionic. Sodium has the structure

$$\text{Na } 1s^2\, 2s^2\, 2p^6\, 3s^1$$

The chlorine atom has the structure

$$\text{Cl } 1s^2\, 2s^2\, 2p^6\, 3s^2\, 3p^5$$

The sodium atom has one unpaired electron in the 3s level and the chlorine one unpaired electron in the 3p level. If an electron is transferred from the sodium to the chlorine then the resulting structures would be

$$(\text{Na } 1s^2\, 2s^2\, 2p^6)^+ \quad \text{and} \quad (\text{Cl } 1s^2\, 2s^2\, 2p^6\, 3s^2\, 3p^6)^-$$

that is, a sodium ion with unit positive charge and a chlorine ion with unit negative charge are produced. Since these ions are oppositely charged the normal electrostatic or Coulombic force of attraction now comes into play and the ions are attracted to each other. Since many ions are formed during a chemical reaction, they will arrange themselves in a regular three-dimensional array or crystal lattice, in order to achieve a state of minimum potential energy. A section of the sodium chloride lattice is shown in Fig. 3.1.

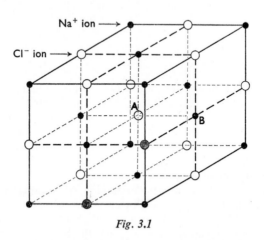

Fig. 3.1

The sodium chloride lattice shows one possible arrangement of ions for an ionic solid. Other compounds can have different lattice arrangements but in all of them the following two conditions are fulfilled:

(i) Each ion has as its nearest neighbours ions of opposite charge. (e.g. Chlorine ion A has six sodium ions as its nearest neighbours, while sodium ion B has six chlorine ions as its nearest neighbours.)
(ii) The solid as a whole is electrically neutral.

A more detailed discussion of the structure of ionic solids is given in Chapter 5.

The above discussion describes *how* an ionic bond is formed but does not reveal *why* sodium and chlorine combine to give an ionic solid. To understand this the energetics of the change must be considered. The overall change can be represented by

$$Na(s) + \tfrac{1}{2}Cl_2(g) \xrightarrow{\Delta H} Na^+Cl^-(s)$$

where s stands for solid and g for gas. ΔH stands for the heat of reaction or enthalpy of reaction. A positive sign for ΔH indicates that heat is absorbed *by* the system *from* the surroundings. A negative sign indicates that heat is given out *by* the system *to* the surroundings. Assuming that all systems tend to a state of minimum potential energy, then sodium chloride will be formed by this reaction if ΔH is negative and large, and provided that no other process can give a larger value. The change can be broken down into a number of stages, thus:

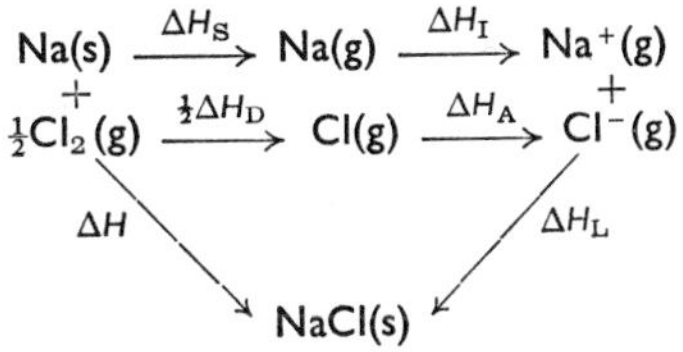

$$Na(s) \xrightarrow{\Delta H_S} Na(g) \xrightarrow{\Delta H_I} Na^+(g)$$
$$+$$
$$\tfrac{1}{2}Cl_2(g) \xrightarrow{\tfrac{1}{2}\Delta H_D} Cl(g) \xrightarrow{\Delta H_A} Cl^-(g)$$

$$\Delta H \searrow \qquad \swarrow \Delta H_L$$

$$NaCl(s)$$

Fig. 3.2

This method of representing the energy changes involved in the formation of a compound is known as a Born–Haber cycle.

Hess's law of constant heat summation states that if a system changes from a state A to a state B the total quantity of heat evolved or absorbed is independent of the method of going from A to B. In this case, therefore, one can write:

$$\Delta H = \Delta H_S + \Delta H_I + \tfrac{1}{2}\Delta H_D + \Delta H_A + \Delta H_L$$

Here ΔH_S is the *heat of sublimation* of sodium, which is positive since energy must be supplied to separate the atoms of the solid; ΔH_I is the *heat of ionization* of sodium or the ionization potential, which is positive since again energy must be supplied in removing the valency electron from a sodium atom; ΔH_D is the *heat of dissociation* of chlorine molecules, which is positive since energy is required to break up the molecules into atoms; ΔH_A is the *electron affinity* of chlorine, which is negative since energy is given out as a chlorine atom attracts an electron; and ΔH_L is the *lattice energy* of sodium chloride, which is negative since energy is evolved as the sodium and chloride ions attract each other to form the crystal lattice.

ΔH will be negative if

$$\Delta H_A + \Delta H_L > \Delta H_S + \Delta H_I + \tfrac{1}{2}\Delta H_D$$

If ΔH is negative then the product is in a state of lower potential energy than the reactants; in other words, the product is more 'stable' than the reactants, and so the reaction may occur. ΔH_I and ΔH_A vary in a regular manner with the position of the element in the Periodic Table. This effect will be discussed later.

The ionization potential and electron affinity factors may be considered from the point of view of electronic structure. If ΔH_I is low and ΔH_A is high, an ionic bond may be expected. The electronic structure of the sodium ion is the same as that of the noble gas neon. The electronic structure of the chlorine ion is the same as that of argon. But neon and argon are both very unreactive elements, so their structure must be relatively stable. This means that if an electron is removed from sodium a very stable electronic structure results; therefore one would expect ΔH_I to be relatively small. If an electron is donated to chlorine once again a very stable electronic structure results; therefore one would expect ΔH_A to be high. This is the basis of the so-called *octet* rule which states that atoms combine together so as to obtain eight electrons in their valency shells. The rule applies without qualification for ionic bonding, only to Periods 2 and 3 of the Periodic Classification. The octet or noble gas structure does represent a stable structure, but it is not the *only* stable structure which is possible. Consider lead(II) chloride, for example. Its formation can be represented as

$$\text{Pb } (2, 8, 18, 32, 18) \; 6s^2 \; 6p^2 \qquad \begin{array}{l} \text{Cl } 1s^2 \; 2s^2 \; 2p^6 \; 3s^2 \; 3p^5 \\ \text{Cl } 1s^2 \; 2s^2 \; 2p^6 \; 3s^2 \; 3p^5 \end{array}$$

$$\downarrow$$

$$[\text{Pb } (2, 8, 18, 32, 18) \; 6s^2]^{2+} \quad 2[\text{Cl } 1s^2 \; 2s^2 \; 2p^6 \; 3s^2 \; 3p^6]^-$$

The lead(II) ion does not have a noble gas structure though all its electrons are paired. For compounds which are almost purely ionic, however, the octet rule has wide application.

There remains one final question to be answered: what prevents the positive and negative ions collapsing into each other, so causing the destruction of the crystal lattice? The answer to this lies in the structure of the ions. Suppose two ions are approaching each other as shown in Fig. 3.3. There is an attractive force because one ion has a

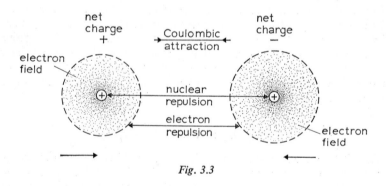

Fig. 3.3

net positive charge and the other a net negative charge. There are also two repulsion terms: repulsion between the positively charged nuclei and repulsion between the negatively charged electron fields. The ions will come to equilibrium when the attractive and repulsive forces balance.

The ions do not actually come to rest, of course, since thermal energy causes oscillation about the equilibrium positions. The attractive and repulsive forces can be summed over the whole lattice for an aggregate of ions and will account for the lattice energy.

The covalent bond

In covalency, bonding arises by the mutual sharing of one or more electrons, rather than by complete transfer of electrons from one atom to another.

The hydrogen molecule ion (H_2^+)

The simplest possible example of covalent bonding is found in the hydrogen molecule ion. There are only three particles involved, two protons and an electron. The forces involved in this molecule are shown schematically in Fig. 3.4. The nuclei approach each other until the inter-nuclear repulsive force is balanced by the force of attraction between

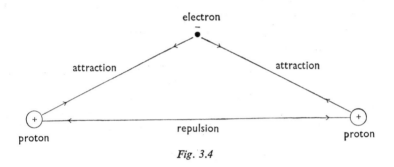

Fig. 3.4

the nuclei and the extra-nuclear electron. This takes into account only potential energy terms. The kinetic energies of the electron and nuclei work against bond formation. The covalent bond, therefore, arises because of the lowering in potential energy when an electron is close to two nuclei. In terms of the charge–cloud model the covalent bond may be represented as an increase in electron density between the two nuclei which arises when overlap of orbitals occurs between two atoms. In this case this could be represented as in Fig. 3.5.

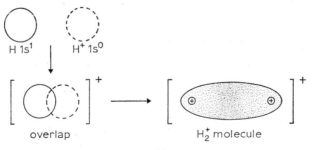

Fig. 3.5

Conditions for bond stability

When any two atoms approach each other orbital overlap can occur to *some extent*, that is, a bond will be formed. A helium atom has the structure

$$He\ 1s^2\ 2s^0\ 2p^0$$

If two helium atoms approach each other then the 1s electron cloud can overlap with the empty 2s or 2p orbitals of the other helium atom. There can be no significant overlap, however, between the 1s orbitals of the two helium atoms. This is because the 1s orbitals contain two paired electrons; that is, they are full. The atoms cannot approach each other closely and so only a weak interaction can occur. Overlap between the valency orbitals of one atom and the extravalency* orbitals of another gives rise to very weak Van der Waals forces (e.g. in solid helium), with a bond energy† less than 17 kJ mol^{-1}.

For more significant overlap and hence more stable bonds there are three possible conditions:

(i) One atom has an empty valency orbital and the other a half-filled one; for example, the H_2^+ molecule discussed above with a bond energy of 270 kJ mol^{-1}.

(ii) Both atoms have half-filled valency orbitals, for example, in the H_2 molecule where initially the hydrogen atom has the structure

$$H\ 1s^1$$

This is the normal covalent bond. The bond energy is 456 kJ mol^{-1}.

(iii) One atom has a filled valency orbital not being used for bonding and the other a vacant valency orbital; for example, $F_3B:NH_3$, which is discussed in detail on page 50. When one atom supplies both electrons for the valency bond this is sometimes known as *coordinate bonding* or *dative covalency*. The bond energy is usually less than 210 kJ mol^{-1}.

Examples of types (ii) and (iii) and of multiple bonds are discussed in detail later.

Homonuclear diatomic molecules

These are molecules made up of two atoms of the same element.

---

* Extravalency orbitals are those vacant orbitals of higher energy than the valency orbitals.

† The energy required to break up water, for example, into its constituent atoms is 925 kJ mol^{-1}. Since water contains two O–H bonds the average energy required to break one of these bonds is $\frac{1}{2}$ (925), that is 462·5 kJ. This is called the *average bond energy* or simply the bond energy of the O–H bond.

The hydrogen molecule H_2

The hydrogen atom has the electronic structure H $1s^1$, so that a covalent bond is formed by the overlap of the two half-filled 1s orbitals. This is shown in Fig. 3.6, where an unshaded area enclosed by a full line

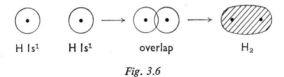

Fig. 3.6

represents a half-filled orbital, and the shaded area represents a full orbital.

The chlorine molecule

The chlorine atom has the structure Cl $1s^2 2s^2 2p^6 3s^2 3p^5$. The formation of the Cl_2 molecule can be represented by Fig. 3.7 (showing only the valency p orbitals).

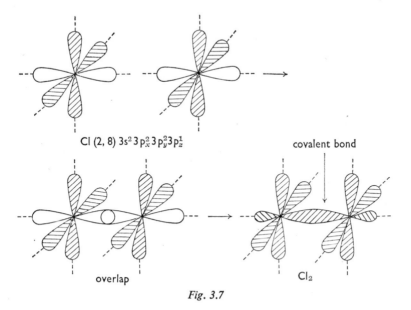

Cl (2, 8) $3s^2 3p_x^2 3p_y^2 3p_z^1$ covalent bond

overlap Cl_2

Fig. 3.7

The representation of the Cl_2 molecule according to this model is not strictly accurate as it does not give the positions of the lone pairs as indicated by other experimental evidence. This will be referred to later (p. 39).

2*

Since the molecules discussed above show bonding between identical atoms, the electrons will be equally shared, so that pure covalency results. The situation for heteronuclear molecules is somewhat different.

Heteronuclear molecules

These are molecules containing atoms of different elements bonded together.

The hydrogen chloride molecule HCl

The hydrogen atom has the structure H $1s^1$. The chlorine atom has the structure Cl $1s^2 2s^2 2p^6 3s^2 3p^5$. The formation of hydrogen chloride, therefore, can be represented by Fig. 3.8. However, the chlorine atom

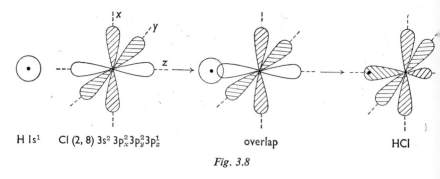

H $1s^1$ Cl $(2, 8)$ $3s^2$ $3p_x^2 3p_y^2 3p_z^1$ overlap   HCl

Fig. 3.8

has a greater tendency to attract electrons than the hydrogen atom, so that the charge density is greater near the chlorine than near the hydrogen. Leaving out the filled non-bonding p orbitals on the chlorine, the hydrogen chloride molecule can be represented by Fig. 3.9.

$$\overset{\delta+}{H}\text{————}\overset{\delta-}{Cl}$$

Fig. 3.9

The hydrogen atom may be considered to possess a fractional positive charge $\delta+$ and the chlorine a fractional negative charge $\delta-$. The bonding in hydrogen chloride may be regarded as intermediate between ionic and covalent bonding rather than purely covalent. In all heteronuclear molecules the different tendency to attract electrons shown by different atoms must be taken into account in describing the final charge density distribution. Since most compounds contain atoms of different elements, intermediate bonding, lying somewhere between purely ionic bonding and pure covalent bonding, is extremely common.

Molecular shape

A major advantage of the charge cloud model is that it can predict approximate molecular shape. The diatomic molecules discussed so far must be linear. For more complex molecules, however, a variety of shapes is possible. Consider the water molecule H_2O. The structure of the hydrogen atom is

$$H\ 1s^1$$

and that of the oxygen atom is

$$O\ 1s^2\ 2s^2\ 2p_x^2\ 2p_y^1\ 2p_z^1$$

The formation of a water molecule could therefore be represented by Fig. 3.10.

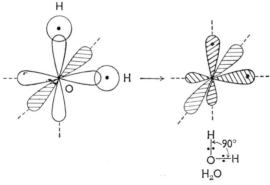

Fig. 3.10

Since the original half-filled p orbitals of the oxygen are at right angles, the expected shape is as shown, that is a bent molecule with an angle of 90° between the two hydrogen–oxygen bonds. However, experiment shows that the bond angle is greater than this (104·5°). A reasonable explanation is that the bonding electron pairs are forced further apart by mutual repulsion.

Consider ammonia as an example of a tetra-atomic molecule. The structure of hydrogen is

$$H\ 1s^1$$

and that of nitrogen is

$$N\ 1s^2\ 2s^2\ 2p_x^1\ 2p_y^1\ 2p_z^1$$

The formation of ammonia, therefore, is represented by Fig. 3.11. It would be expected to be a pyramidal molecule with bond angles of

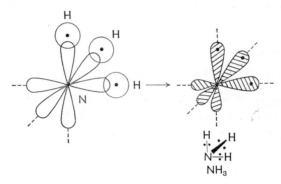

Fig. 3.11

approximately 90°. In fact the bond angles are considerably greater than 90° (107·5°), which is again explained by the mutual repulsion of the bonding pairs.

Hybridization of orbitals

sp³ hybridization

Although the method of the last section gives the approximate shapes of molecules, it has one important defect—it does not correctly indicate the positions of the lone (non-bonding) pairs in the molecules. Again, the discussion (so far) implies that the valency of an element is equal to the number of unpaired electrons in the valency shell; for example, oxygen has the structure

$$O\ 1s^2\ 2s^2\ 2p_x^2\ 2p_y^1\ 2p_z^1$$

It has two unpaired electrons and hence a valency of 2. But what about carbon? Carbon has the structure

$$C\ 1s^2\ 2s^2\ 2p_x^1\ 2p_y^1\ 2p_z^0$$

It has two unpaired electrons and hence might be expected to be divalent. However, carbon shows a covalency of four. Both of these points can be satisfactorily explained by a further elaboration of the orbital model. Consider the carbon atom. The structure of carbon *in the ground state* is certainly given by the orbital theory as shown above, corresponding to the structure shown in Fig. 3.12 (a dotted, unshaded lobe represents an *empty* orbital). There are, however, other possible

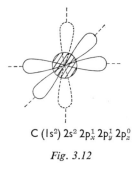

C $(1s^2)\ 2s^2\ 2p_x^1\ 2p_y^1\ 2p_z^0$

Fig. 3.12

orientations for the valency electron charge clouds. Suppose the following process occurs:

(1) The 2s electrons unpair and one is promoted to the vacant $2p_z$ orbital, giving

$$C\ 1s^2\ 2s^1\ 2p_x^1\ 2p_y^1\ 2p_z^1$$

(2) The 2s and three 2p orbitals are then replaced by four equivalent hybrid orbitals. These will be orientated towards the vertices of a regular tetrahedron as shown in Fig. 3.13.

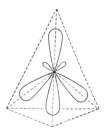

Fig. 3.13   Orientation of four equivalent sp³ hybrid orbitals

This hypothetical process may be used to account for the tetrahedral arrangement of hydrogen atoms in methane. Since the four equivalent orbitals are formed from one s orbital and three p orbitals, this is known as sp³ hybridization. In methane, for example, the four half-filled sp³ hybrid orbitals overlap with the half-filled s orbitals of four hydrogen atoms to give the tetrahedral methane molecule, as shown in

Fig. 3.14. (For simplicity, future diagrams will omit small lobes.) In the proposed process:

one isolated   four isolated
carbon atom + hydrogen atoms $\longrightarrow$ a methane molecule

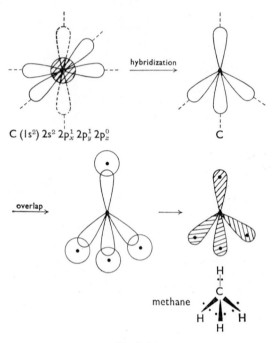

C $(1s^2)\ 2s^2\ 2p_x^1\ 2p_y^1\ 2p_z^0$

hybridization

C

overlap

methane

Fig. 3.14

the promotion of a 2s electron requires energy and so is energetically unfavoured. This is offset, however, by the additional lowering in energy of the system by

(*a*) four hydrogen s orbitals overlapping instead of two (if carbon were to remain with only two unpaired electrons), and

(*b*) the lowering of potential energy when the molecule adopts a tetra-hedral configuration, keeping mutually repelling bonding pairs as far away from each other as possible.

The shapes of the water and ammonia molecules may be interpreted in terms of sp³ hybridization. Oxygen has the structure

$$O\ 1s^2\ 2s^2\ 2p_x^2\ 2p_y^1\ 2p_z^1$$

If sp^3 hybridization occurs this gives two filled sp^3 orbitals and two half-filled sp^3 orbitals. The two half-filled orbitals then overlap with the half-filled 1s orbitals of hydrogen to give water, as shown in Fig. 3.15.

The bond angle for a regular tetrahedron is 109° 28′, while that for H–O bonds in water is 105° (approximately). There is, therefore, some deviation from the ideal. This is accounted for by the properties of the lone pairs. The lone pair electrons are attracted by only one nucleus and so are much 'freer' than the highly localized bonding pairs. Consequently the lone pair electron clouds tend to expand to fill as much space

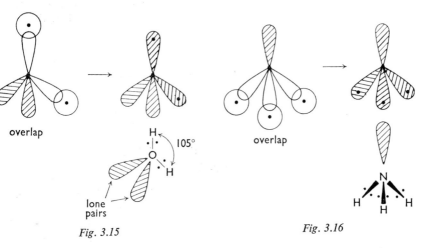

overlap

lone
pairs

Fig. 3.15

overlap

Fig. 3.16

as possible and in so doing repel the bonding electrons, pushing them closer together. The result is that the bond angle is less than 109°.

Nitrogen has the structure

$$N\ 1s^2\ 2s^2\ 2p_x^1\ 2p_y^1\ 2p_z^1$$

sp^3 hybridization gives one filled sp^3 orbital and three half-filled sp^3 orbitals. Overlap then occurs with these half-filled orbitals and the half-filled 1s orbitals of three hydrogen atoms, as shown in Fig. 3.16. The molecule is, therefore, pyramidal with the lone pair in the unoccupied tetrahedral position. Once again the bond angles are less than 109° because of the repulsive effect of the lone pair.

In the descriptions of the ammonia and water molecules given above, the predicted lone pair positions are more in accordance with experimental data than the positions predicted by the non-hybridization model (p. 35).

sp^2 hybridization

Hydridization of two p orbitals and one s orbital, giving three equivalent sp² orbitals, is a way of describing some molecules. In the ground state, boron has the structure

$$\text{B } 1s^2 \, 2s^2 \, 2p_x^1 \, 2p_y^0 \, 2p_z^0$$

sp² hybridization may be represented by the following steps:

(1) Unpairing and promotion of a 2s electron to a vacant 2p level to give

$$\text{B } 1s^2 \, 2s^1 \, 2p_x^1 \, 2p_y^1 \, 2p_z^0$$

(2) Combination of the 2s orbital with the two p orbitals to give three equivalent sp² hybrid orbitals lying in one plane and pointing towards the vertices of an equilateral triangle, as in Fig. 3.17. (The three small lobes are left out of future diagrams for simplicity.)

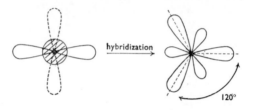

three sp² hybrid orbitals

Fig. 3.17

The trigonal boron trifluoride molecule would then be formed by overlap with the half-filled p orbitals of three fluorine atoms. The valency electrons of the fluorine are in sp³ orbitals, three filled and one half-filled. The formation of boron trifluoride may be represented by Fig. 3.18 (leaving out the lone pairs on the fluorine atoms).

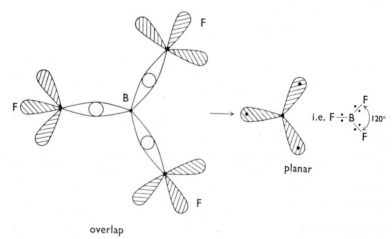

overlap

Fig. 3.18

sp hybridization

This can be thought of in the following terms:

(1) Unpairing of 2s electrons and promotion of one to a vacant p orbital.
(2) Hybridization to give two equivalent sp hybrid orbitals arranged in a straight line.

Beryllium has the structure

$$Be \; 1s^2 \; 2s^2 \; 2p_x^0 \; 2p_y^0 \; 2p_z^0$$

Promotion gives

$$Be \; 1s^2 \; 2s^1 \; 2p_x^1 \; 2p_y^0 \; 2p_z^0$$

Hybridization gives two equivalent orbitals, as shown in Fig. 3.19.

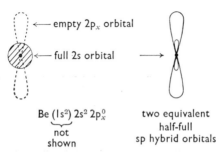

← empty $2p_x$ orbital

← full 2s orbital

Be $(1s^2)$ $2s^2$ $2p_x^0$
not
shown

two equivalent
half-full
sp hybrid orbitals

Fig. 3.19

(Again, the two small lobes will be omitted from future diagrams.) An empty orbital is represented by a dotted line. Beryllium chloride could be represented by Fig. 3.20 (non-bonding electron pairs on the chlorine atoms being omitted).

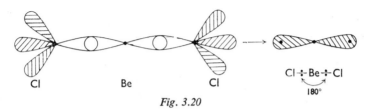

Cl — Be — Cl
180°

Cl   Be   Cl

Fig. 3.20

To summarize: hybridization is the combination of atomic orbitals to give new orbitals of different shape and orientation. Thus orbital theory is used to account for the known geometry of molecules and as a pointer to the geometry of less well-known molecules. Although promotion of an electron occurs in the formal processes discussed above, this is not essential to hybridization. The hybrid orbitals are the result of

Constituent orbitals	Name of hybrid	Geometry*
One s, three p	sp^3	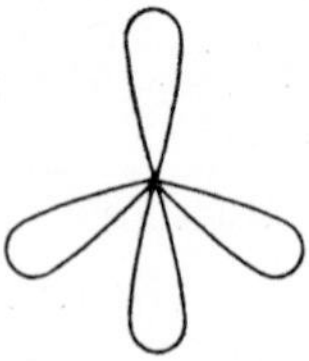 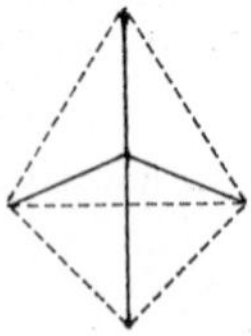 tetrahedral
One s, two p	sp^2	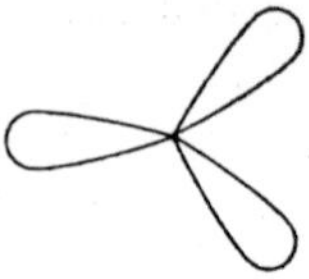 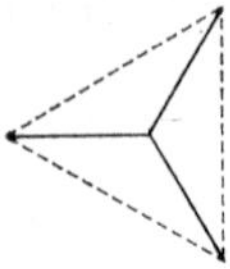 trigonal
One s, one p	sp	  linear
One s, three p, one d	sp^3d	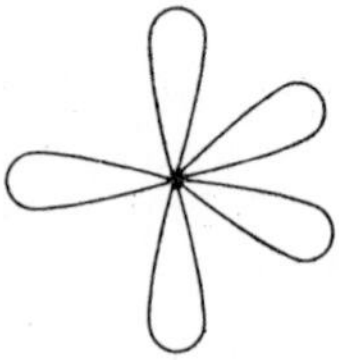 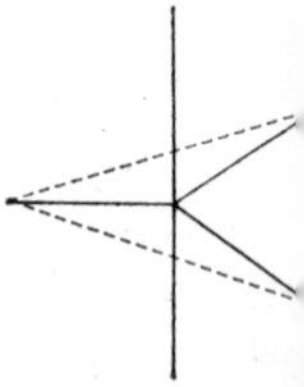 trigonal bi-pyramidal
One s, three p, two d	sp^3d^2	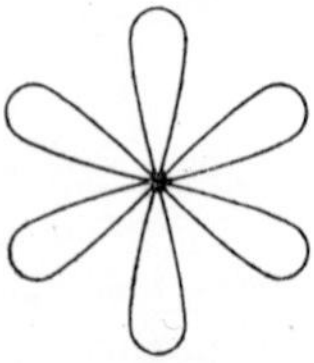 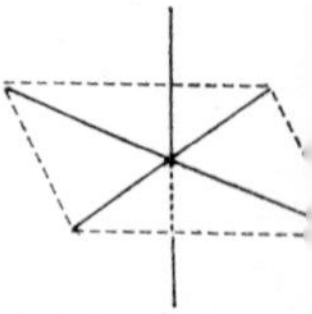 octahedral

* Small lobes of probability are omitted.

Fig. 3.21

combining the wave functions for the 'constituent' orbitals according to the results of wave mechanics. This formal process could equally well be used for empty, partly filled or completely filled orbitals. A summary of the most common types of hybridization and the orientation of the resulting hybrid orbitals is shown in Fig. 3.21.

Multiple bonds

So far, molecules containing only single covalent bonds have been considered. The charge cloud model can also be used to give a satisfactory account of double and triple bonds consistent with their properties.

Ethylene has the classical structure

$$
\begin{array}{c}
H \qquad\qquad H \\
\diagdown\qquad\diagup \\
C{=}C \\
\diagup\qquad\diagdown \\
H \qquad\qquad H
\end{array}
$$

The bond angles are 120° and the molecule is planar. Carbon has the structure

$$C\ 1s^2\ 2s^2\ 2p_x^1\ 2p_y^1\ 2p_z^0$$

Hydrogen has the structure

$$H\ 1s^1$$

If the carbon undergoes sp^2 hybridization this gives the charge cloud arrangement shown in Fig. 3.22a. If overlap now occurs between an sp^2

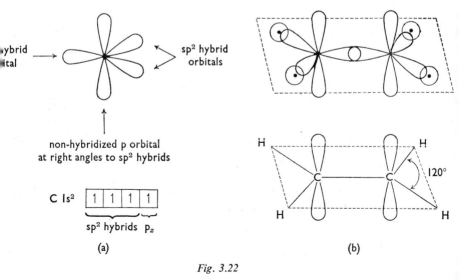

Fig. 3.22

hybrid orbital on each of two carbon atoms and the hydrogens are bonded in the usual way by overlap with the remaining sp^2 hybrids, this gives the structure shown in Fig. 3.22b.

The second carbon–carbon bond can then be formed only by *sideways* overlap of the non-hybridized p_z orbitals, to give a banana-shaped charge cloud above and below the plane of the C_2H_4 skeleton, between the two carbon atoms, as shown in Fig. 3.23.

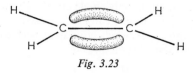

Fig. 3.23

The bonds between the carbons in ethylene, therefore, are different in character. The first bond formed by direct, 'head-on' overlap of two sp^2 hybrid orbitals is called a σ-bond. The charge density is uniformly distributed about the axis of the bond. The second bond, formed by 'sideways' overlap, is called a π-bond. The charge density is not uniformly distributed about the axis of the bond. Fig. 3.24 shows the

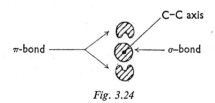

Fig. 3.24

two types of bond in ethylene, *looking along the axis* of the C–C bond.

The *two* charge clouds of the π-bond represent only *one* bond, i.e. two shared electrons. Because the π-bonded electrons are spread out over a greater volume, the charge density of the π-bond will be relatively lower than that of a σ-bond and so it will be more readily polarized (see p. 243). Again, the π-bond inhibits free rotation of the CH_2 groups about the carbon–carbon axis. For rotation to occur, the degree of sideways overlap between the p_z orbitals must be decreased. This requires energy, so free rotation is inhibited. Rotation about a σ-bond can occur quite freely because it does not alter the degree of overlap, since the charge density is uniformly distributed about the bond axis.

Triple bonds are formed in a similar manner to double bonds. Consider acetylene. Carbon has the structure

$$C\ 1s^2\ 2s^2\ 2p_x^1\ 2p_y^1\ 2p_z^0$$

Hydrogen has the structure

$$H\ 1s^1$$

If the carbon undergoes sp hybridization this gives the structure shown in Fig. 3.25a. Overlap can then occur, giving one carbon–carbon σ-bond and two hydrogen–carbon σ-bonds, as shown in Fig. 3.25b. Sideways

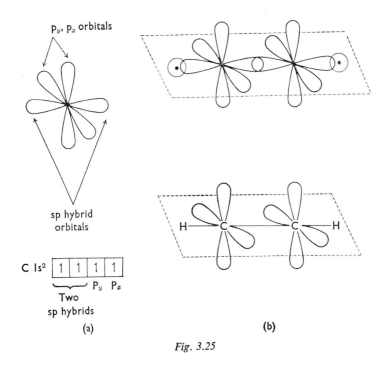

Fig. 3.25

overlap of the non-hybridized p orbitals then gives two π-bonds, as shown in Fig. 3.26.

In some molecules 'sideways' overlap occurs between p orbitals on

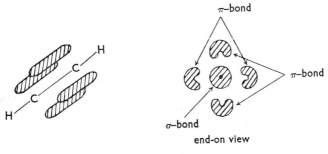

Fig. 3.26

more than two nuclei. Consider, for example, carbon dioxide. Carbon has the structure

$$\text{C } 1s^2 \, 2s^2 \, 2p_x^1 \, 2p_y^1 \, 2p_z^0$$

and oxygen the structure

$$\text{O } 1s^2 \, 2s^2 \, 2p_x^2 \, 2p_y^1 \, 2p_z^1$$

If both the carbon and the oxygen undergo sp hybridization, this gives the structures shown in Fig. 3.27a and b.

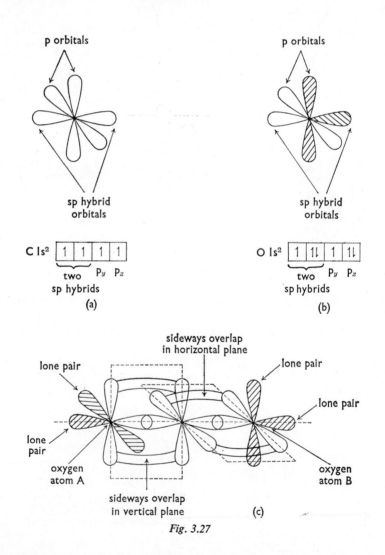

Fig. 3.27

The linear structure of carbon dioxide can then be represented schematically by Fig. 3.27c. However, there is no reason why the p orbital lone pair on oxygen atom A should not be in the vertical position rather than the horizontal, or why the p orbital lone pair on oxygen B should not be in the horizontal position. Sideways overlap therefore occurs between all the p orbitals in the vertical plane and all those in the horizontal plane. There are four electrons per three vertical p orbitals and four electrons per three horizontal orbitals. The π-bonds are said to be delocalized, as shown in Fig. 3.28.

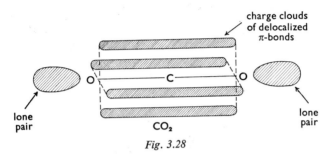

Fig. 3.28

Delocalization is an important phenomenon, especially in organic compounds such as benzene. Benzene has the classical Kekulé structure

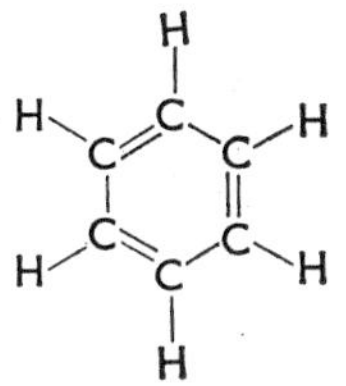

Fig. 3.29

shown in Fig. 3.29. Carbon has the structure

$$\text{C } 1s^2 \ 2s^2 \ 2p_x^1 \ 2p_y^1 \ 2p_z^0$$

and hydrogen the structure

$$\text{H } 1s^1$$

If the carbon undergoes sp^2 hybridization the C_6H_6 skeleton is formed as shown in Fig. 3.30. This leaves one electron on each carbon in a p orbital, as shown in Fig. 3.31. (The hydrogen atoms are not shown.)

The alternate double bonds may now be formed by π overlap between the p orbitals of atoms a and b, c and d and e and f, or by overlap between atoms b and c, d and e and f and a. Both these structures are

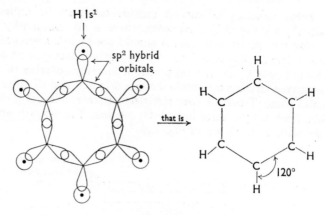

Fig. 3.30

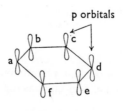

Fig. 3.31

equi-probable, so delocalization occurs and overlap takes place around the whole perimeter of the ring, to give 'bicycle tyres' of charge cloud above and below the plane of the ring, as shown in Fig. 3.32. This is the

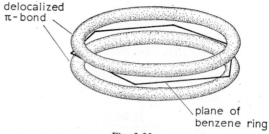

Fig. 3.32

accepted charge cloud structure of benzene. The evidence in support of this structure is given in Chapter 11. It is worth pointing out now, however, that benzene is more stable than the Kekulé structure would lead one to expect. Delocalization gives additional stability to the molecule.

If the charge cloud is regarded as an 'electron gas' then increase in volume (by delocalization) lowers the potential energy of the system, so increasing the stability.

A discussion of the structure of benzene is convenient for illustrating two possible approaches to bonding. The approach used above considers the electron distribution with respect to the molecule as a whole. The other possible approach is an extension of classical valency theory. Consider once again, for example, the structure of benzene. According to classical theory there are at least two possible structures for benzene, as shown in Fig. 3.33. The actual structure cannot be represented on

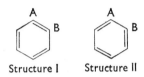

Structure I   Structure II

Fig. 3.33

paper by either I or II but is intermediate and must be represented by both structures written together. The actual structure is said to be a *resonance hybrid* of structures I and II, which are called *canonical structures*. (Note that the term 'resonance hybrid' has nothing to do with orbital hybridization or oscillation of bonds.) The difference between the predicted energy of formation of structure I or II and the actual energy of formation of benzene is called the *resonance energy*.

The two approaches to the problem of bonding are equivalent, but are expressed in different terms. Occasionally, a compromise language is used; for example, when the structure of benzene is represented as

the dotted line representing the delocalized π-bond.

The coordinate or dative bond

A third possible way in which combination can occur, other than by complete electron transfer or mutual electron sharing, is by the overlap of an orbital containing an electron pair on one atom, with an empty orbital on another. Ammonia, for example, possesses a lone pair, while in boron trifluoride the boron atom has an empty p orbital as shown in Fig. 3.34a and b.

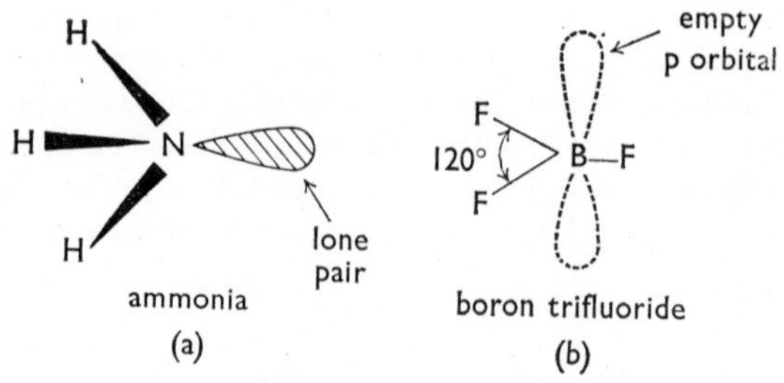

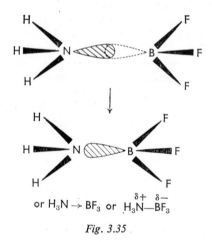

H₃N

ammonia

(a)

empty
p orbital

boron trifluoride

(b)

Fig. 3.34

If sp³ hybridization is assumed then the three boron–fluorine bonds become tetrahedrally orientated with an empty sp³ hybrid orbital in the fourth tetrahedral position. Overlap can then occur as shown in Fig. 3.35. The arrowed bond indicates that both bonding electrons are sup-

or $H_3N \rightarrow BF_3$ or $H_3\overset{\delta+}{N}-\overset{\delta-}{BF_3}$

Fig. 3.35

plied by the nitrogen. This type of bond is called a *coordinate bond*. Because one atom provides both the electrons for sharing, there is often a fractional positive charge on the donor atom and a fractional negative charge on the acceptor atom. Care must be taken in predicting charge distribution, since this is influenced by other factors such as the relative electro-negativities of the atoms making up the molecule. It is not usual to distinguish between coordinate covalency and ordinary covalency since in both cases the bond consists of a pair of mutually shared electrons. Even the fractional charges do not distinguish between them, since these are found when atoms of differing electron attraction are covalently bonded. The bond energy for a coordinate bond is in the order of 210 kJ mol^{-1}.

Metallic bonding

This is a type of bonding peculiar to the metallic state. Metals possess certain characteristic properties such as relatively high electrical and thermal conductivities, malleability and ductility, which suggest a distinctive type of bonding. Elements showing metallic characteristics possess the following properties:

(1) low first ionization potentials (below 1050 kJ mol^{-1}),
(2) few valency electrons relative to the number of valency orbitals.

For example, sodium has the structure

$$Na\ 1s^2\ 2s^2\ 2p^6\ 3s^1\ 3p^0\ 3d^0$$

It has an ionization potential (for the removal of the 3s valency electron) of 494 kJ/mol^{-1}. Sodium atoms, therefore, cluster together closely so that the few available valency electrons occupy as many as possible of the empty valency orbitals. The valency electrons are not localized between particular atoms but move about freely and randomly throughout the whole metallic crystal. The sodium metal crystal, therefore, behaves as if it were a regular array of positively charged sodium ions immersed in a 'sea' of randomly moving electrons. This simple model can account for many of the characteristic properties of metals. A more detailed account of metallic bonding and its relationship to properties will be given in Chapter 6.

Hydrogen bonding

Although water has the molecular structure shown in Fig. 3.36, its physical properties indicate that it is associated in the liquid state. At

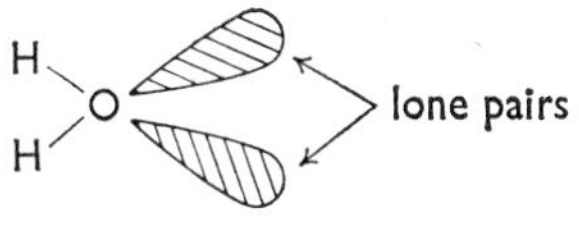

Fig. 3.36

first sight it is difficult to see how association can occur, since the hydrogen atom is saturated when it has formed a single covalent link (the first main energy level then contains two electrons) and oxygen is saturated when it has formed two covalent links (the second main energy level then contains eight electrons).

Association is accounted for by the formation of a unique type of bond known as the *hydrogen bond*. Oxygen attracts electrons strongly, compared to hydrogen, so that the water molecule should be represented as shown in Fig. 3.37. Electrons are pulled towards the oxygen, leaving

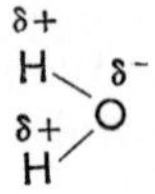

Fig. 3.37

a partially exposed proton at the other end of the bond, since hydrogen has no 'inner' electrons. The positively charged hydrogen then interacts electrostatically with the electron field of the oxygen of another water molecule, giving a hydrogen bond. The bond is much weaker than the covalent or ionic bond, having a bond energy in the range $12 \cdot 5$–$30 \, \text{kJ mol}^{-1}$. It is usual to represent it by a dotted line, so that associated water may be represented by Fig. 3.38.

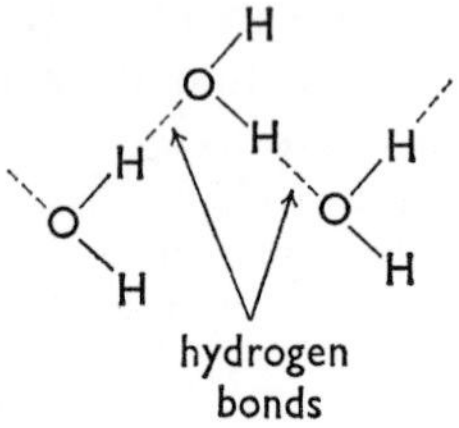

hydrogen
bonds

Fig. 3.38

Hydrogen bonds are formed only when the hydrogen is bonded to an element which strongly attracts electrons (i.e. an element with relatively high electronegativity and small atomic size), e.g. when it is bonded to oxygen, nitrogen or fluorine. They are very important in biological systems; for instance, in proteins.

Van der Waals bonds

All gases at sufficiently low temperatures condense to liquids and eventually solids. If the molecule of the gas possesses a permanent dipole, as does hydrogen chloride, for example, then it is easy to see why this should occur. The dipoles interact with each other, causing the molecules to be attracted to each other. Random thermal motion works against this attraction so that only when the temperature is lowered sufficiently does a condensed phase form. But what of gases such as helium, which have atoms with no incomplete valency orbitals and no permanent dipole? The Van der Waals attraction for this type of atom has been explained by the London 'instantaneous dipole' theory. The orbital theory gives a *time average* picture of the helium atom as shown in Fig. 3.39a. The probability of finding an electron in a particular ele-

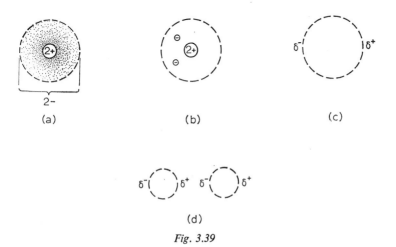

(a)   (b)   (c)

(d)

Fig. 3.39

ment of volume is spherically symmetrical about the nucleus. At a *particular instant*, however, the electrons may be in the positions shown in Fig. 3.39b. This amounts to the atom having an instantaneous dipole as shown in Fig. 3.39c. This instantaneous dipole induces an energetically favourable dipole in neighbouring atoms, as shown in Fig. 3.39d, so that the atoms attract each other. For helium the Van der Waals interaction energy is 0.084 kJ mol^{-1}. In general it is of the order of 4.2 kJ mol^{-1} (assuming no *permanent* dipole).

References

Books

A Structural Introduction to Chemistry, E. T. Harris (Blackie)
Valency and Molecular Structure, Cartmell and Fowles (Butterworths)
Chemistry: An Experimental Science, CHEM Study, Pupil's Book and
    Teacher's Guide (W. H. Freeman & Co.)
Chemical Systems, CBA project (McGraw-Hill)

Structural Principles in Inorganic Compounds, W. E. Addison (Longmans)

An Introduction to Modern Chemistry, M. J. S. Dewar (Athlone Press)

Films

Chemical Bonding, G. C. Pimentel

Shapes and Polarities of Molecules, D. Dows

Both of these are CHEM Study Films. They are available from Sound-Services Ltd., cat. nos. 4157/999 and 4154/999 respectively.

4. Bonding and periodicity

Bond type as a function of attraction for electrons

It is implicit in the discussion of bonding in Chapter 3 that the type of bond formed is related to the degree of attraction the bonded atoms have for electrons. Fig. 4.1 shows the situation for purely ionic and purely covalent bonding, and for bonding intermediate between these two.

$$\boxed{A\cdot + B\cdot \to A^+ B:^-} \qquad \boxed{A\cdot + B\cdot \to A:B}$$

$$A^+B^- \longrightarrow \overset{\delta+ \;\; \delta-}{A\text{---}B} \longrightarrow A\text{---}B$$

purely ionic   intermediate   purely covalent

Fig. 4.1

If B has a very much greater attraction for electrons than A, ionic bonding appears the most likely. If the attraction for electrons shown by A is equal to that shown by B, then purely covalent bonding is the most likely. If the attraction for electrons shown by A is different from that shown by B, but not markedly so, intermediate bonding is likely. Therefore, if to each element can be assigned a number which is a quantitative measure of its tendency to attract electrons, then this will help one to predict the most probable bond type when two elements combine. Fortunately, such measures of electron attraction can be obtained, and they are related in a fairly regular way to the position of the element in the Periodic Classification.

Measures of electron attraction

Ionization potential*

This has been mentioned already in the discussion of ionic bonding given in Chapter 3. It may be defined formally as the energy required to

---

* Two important methods of measuring ionization potential are clearly described in the CHEM Study Film *Ionization Energy*, B. H. Mahan.

remove an electron from an isolated atom of an element in the gaseous state to an infinite distance from the nucleus; i.e. the energy absorbed in the following process:

$$X \text{ (gas)} \longrightarrow X^+ \text{ (gas)} + e^- \text{ (gas)}$$

The first ionization potential is the minimum energy required to remove the least tightly held electron. The energy required to remove the next electron, i.e. for the process

$$X^+ \text{ (gas)} \longrightarrow X^{2+} \text{ (gas)} + e^- \text{ (gas)}$$

is called the *second* ionization potential, and so on. When an electron has been removed the atom acquires a net positive charge so that the next electron is more difficult to remove; i.e.

first ionization potential < second ionization potential

and in general,

first I.P. < second I.P. < third I.P.   etc.

Factors affecting the magnitude of the ionization potential

(*a*) *Nuclear charge*

Since according to Coulomb's Law the force of attraction between two charges at a distance r from each other is given by

$$F = k \frac{e_1 e_2}{r^2}$$

where e_1 = magnitude of one charge
e_2 = magnitude of the other charge
k = a constant
r = the distance between the charges

then, for a given electron, the force on it must increase as the nuclear charge increases. This means that as the nuclear charge increases, the energy required to remove the electron becomes progressively greater. Therefore, ignoring other factors, the ionization potential will increase as the nuclear charge (atomic number) increases.

(*b*) *The principal quantum number* n

The energy diagram on p. 20 indicates that, in general, the energy of an electron increases as the principal quantum number n increases. Thus an electron of principal quantum number 1 will be in a state of lower energy than one with principal quantum number 2. This in turn means that it is *harder to remove* an electron for which $n=1$ than one for which $n=2$. As n increases the ionization potential decreases, if other factors are ignored.

(c) The shielding effect of inner shells

In most atoms there are completed or partially completed shells between the outer electron and the nucleus. These inner shells will 'shield' the outer electron from the full effect of the nucleus by exerting a repulsive force on the outer electron. The structure of the inner shells will, therefore, affect the ionization potential.

If other factors remain constant, the ionization potential should decrease as the number of inner electrons increases. However, the shielding effect depends not only on the number of inner electrons but also on the shapes of the electron charge clouds, i.e. on whether the electrons are s, p, d or f. Because of the shapes of the s, p, d and f charge clouds it follows that the average electron density *near the nucleus* decreases in the order

$$s > p > d > f$$

The screening effect of inner electrons also decreases in this order, since a given outer electron will penetrate inner orbitals to a progressively greater extent, in the order f, d, p, s. This penetration effect is further reinforced for completed inner sub-levels by there being 2, 6, 10 and 14 electrons in the s, p, d and f shells. Thus, when an outer electron is screened by a completed d shell, this will be less effective than being screened by a completed p shell because (i) the d orbitals are more easily penetrated because of their shape, and (ii) the outer electron penetrates 10 electrons in the d shell and only 6 in the p shell. This means that if other factors are constant an electron underlain by a completed p shell will have a lower ionization potential than one underlain by a completed d shell.

(d) The sub-level quantum number of the electron being removed

It was pointed out in section *(c)* that the average electron density near the nucleus decreases in the order

$$s > p > d > f$$

Therefore, if the electron being removed is an s electron, it will penetrate inner shells more deeply than if it were a p electron. Similarly, p electrons penetrate more deeply than d, and d more deeply than f, for a given value of *n*. The ionization potential of the outer electron, therefore, decreases for a given value of *n*, in the order

$$s > p > d > f$$

Periodic variation of ionization potential

Since ionization potentials are affected by the structural factors discussed in the last section, it is not surprising that they show a periodic

3+

variation as the atomic number of the element considered, changes. A graph of first ionization potential against atomic number is of the form shown in Fig. 4.2. This typical 'saw-tooth' graph can be interpreted in terms of the factors discussed in the last section.

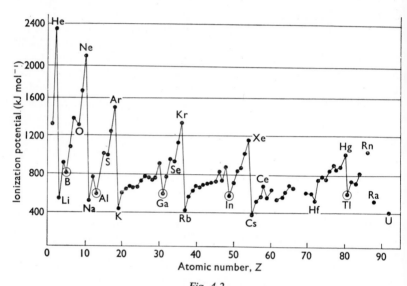

Fig. 4.2

(From *Principles of Atomic Orbitals*, N. N. Greenwood, R.I.C. Monographs for Teachers.

1. Variation of ionization potential for a group

Fig. 4.3 shows graphs of ionization potential against atomic number for groups IM to VIIIM.

The first ionization potential decreases going down a normal group. The factors coming into play in this situation are:

(1) Increase in nuclear charge, which tends to increase the ionization potential.
(2) Increase in principal quantum number n, which tends to decrease the ionization potential.
(3) Increase in number of shells of inner electrons, and hence an increase in the shielding effect of the inner electrons, which tends to decrease the ionization potential.
(4) The sub-level quantum number of the electron being removed is the same for each group, so this factor remains constant.

Factors (2) and (3) outweigh factor (1), so that in general, for a normal group, the first ionization potential decreases going down the group.

For each group there is a relatively steep drop of ionization potential between the first three elements of the group but then the curve begins to flatten out. This can be explained by two effects:

(*a*) In Period 2 the only shielding electrons are two 1s electrons. Although the completed 1s level is a relatively efficient screener, in all other cases the outer electron is shielded by at least eight inner electrons, two of which are in a completed 1s shell.

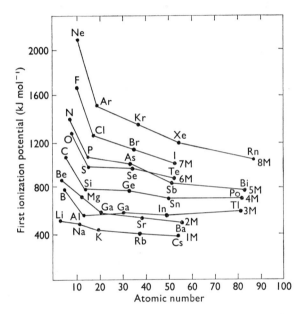

Fig. 4.3 Graph of ionization potential against atomic number (main group elements only)

(*b*) After Period 4 the nuclear charge factor becomes increasingly important because the presence of a transition series in Periods four, five and six considerably increases the magnitude of the nuclear charge. The additional electrons in inner shells, which tend to compensate for this, are d or f electrons which have a relatively poor screening effect. Consequently, after Period 4 the ionization potential decreases more gradually.

For elements of atomic number greater than 71 this tendency is reversed. For an element of atomic number 72 or greater, the nuclear charge is 32 units greater than the next lightest element in the group. The nuclear charge then becomes the important factor, so that the ionization potential starts to increase.

2. Variation of ionization potential for a period

Consider Period 2 elements. The ionization potential plot for these elements has the 'saw-tooth' shape which is characteristic of the nontransition elements in other periods. The plot is shown in Fig. 4.4. The

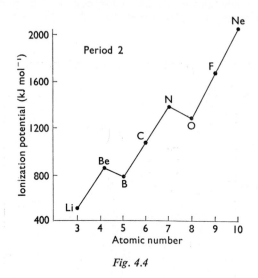

Fig. 4.4

graph indicates that there is a general increase in ionization potential going across the period. But why are there decreases between beryllium and boron, and nitrogen and oxygen? With beryllium a 2s sub-level is completed, which represents a relatively stable configuration. Boron possesses an additional unpaired 2p electron. It is relatively easier to remove this 2p electron than to remove a paired 2s electron from beryllium. The first ionization potential of boron is, therefore, less than that of beryllium.

Nitrogen has the structure

$$1s^2 \, 2s^2 \, 2p_x^1 \, 2p_y^1 \, 2p_z^1$$

Each p orbital contains one unpaired electron. This is a particularly symmetrical distribution and so a relatively stable state. With oxygen this high symmetry is lost by the addition of an extra 2p electron. The first ionization potential of oxygen is, therefore, less than that of nitrogen. The shapes of the graphs for other normal elements of Periods 3, 4, 5 and 6 can be explained in a similar manner.

3. Variation of ionization potential in the d-block elements

The variation of ionization potential for the first transition series is shown in Fig. 4.5. In general, the first ionization potential increases slowly across the series. This is because of the relatively poor screening effect of d electrons; the important factor is the increase in nuclear charge. There is a relatively sharp, though small, increase in ionization potential for manganese. Manganese has the structure $3d^5\ 4s^2$ for its outer two sub-shells. The d level is exactly half-full and the s level full.

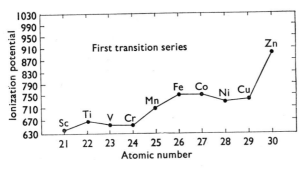

Fig. 4.5

This is a relatively stable arrangement, so the ionization potential is relatively high. The increases in ionization potential of copper and zinc are similarly explained since these also have relatively stable arrangements for the two outer sub-shells, as follows:

Cu: $3d^{10}\ 2s^1$, 3d sub-shell full, 2s sub-shell half-full
Zn: $3d^{10}\ 2s^2$, both outer sub-shells full

Summary In general, the first ionization potentials of the elements show a tendency to increase across the Periodic Table from left to right and a tendency to decrease from top to bottom of the table. For elements with atomic number greater than 71 the latter tendency is reversed. The exceptions to these general trends are discussed in sections (1), (2) and (3) above.

Other measures of electron attraction

Electron affinity

This is defined as the energy evolved in the change

$$X \text{ (gas)} + e^- \text{ (gas)} \longrightarrow X^- \text{ (gas)}$$

This is the first electron affinity, which is the negative of the first ionization energy of the appropriate negative ion, since the process

$$X^- \text{ (gas)} - e^- \longrightarrow X \text{ (gas)}$$

is the reverse of the above process. The second electron affinity can be defined as the energy evolved in the change

$$X^- \text{(gas)} + e^- \text{(gas)} \longrightarrow X^{-2} \text{(gas)}$$

and so on. The factors affecting electron affinities are similar to those which influence ionization potential. As might be expected, the electron affinities tend to decrease going down a normal group and to increase going across the Periodic Table from left to right. There are minor variations from this general rule which can be explained in a similar manner to the variations for ionization potentials. Electron affinities are usually expressed in kJ mol^{-1}. Some electron affinities are shown in Table 4.1.

H 67·4						
Li 52·1	Be —	B 28·4	C 109	N 19·3	O 143	F 349
Na 72	Mg —	Al 38·6	Si 183	P 77·2	S 200	Cl 368
						Br 341
						I 312

Table 4.1   Electron affinities (kJ mol^{-1})

The values for the halogen group show that the electron affinities decrease going down the group from chlorine to iodine, but fluorine is anomalously low. This is the case for all the elements of Period 2, and is probably due to the relatively small size of the atoms of elements in this period, which results in a relatively high repulsive effect on an incoming electron. Electron affinities are difficult to measure experimentally, which detracts considerably from their potential usefulness.

Electronegativity

It was pointed out in Chapter 3 that in heteronuclear molecules, the electrons involved in a covalent bond are not equally shared because the atoms of the two different elements do not attract electrons to an equal extent. This gives rise to a fractional charge on each atom, the molecule as a whole possessing an electric dipole. The *electronegativity* of an atom is a measure of the attraction of that atom for an electron pair present in a covalent bond between that atom and another.

It might be expected that for a molecule X–Y the bond energy would be the average of the bond energies for the molecules X_2 and Y_2. However, it is always the case that

$$E_{XY} > \frac{E_{X_2} + E_{Y_2}}{2}$$

The excess energy arises because of the unequal sharing of the bonding pair in XY, the actual structure of XY being better represented as a resonance hybrid of the canonical structures

$$X—Y \quad \text{and} \quad X^+Y^-$$

Pauling suggested that this excess energy should be fixed by the difference in electronegativities of the two atoms present. He assumed a square relationship and found that the geometric mean was better than the arithmetic mean, because it avoided negative values for electronegativity. This gave the relation

$$(x_X - x_Y)^2 = E_{XY} - \sqrt{(E_{X_2} E_{Y_2})}$$

where x_X = electronegativity of atom X
x_Y = electronegativity of atom Y.

Using this expression Pauling devised a scale of electronegativities which is self-consistent and fits in well with the known facts of descriptive chemistry. Table 4.2 shows some of the electronegativity values from Pauling's scale.

H 2·1						

Li 1·0	Be 1·5	B 2·0	C 2·5	N 3·0	O 3·5	F 4·0
Na 0·9	Mg 1·2	Al 1·5	Si 1·8	P 2·1	S 2·5	Cl 3·0
K 0·8	Ca 1·0				Se 2·4	Br 2·8
Rb 0·8	Sr 1·0				Te 2·1	I 2·5

Table 4.2 Electronegativities (Pauling)

The general trends in electronegativity for main group elements are in agreement with the trends in ionization potential and electron affinity.

The electronegativity decreases going down a group and increases going across the Periodic Table from left to right. Consequently, the most electronegative elements are at the top right of the table and the least electronegative at the bottom left.

The concept of electronegativity must be applied with caution to molecules containing more than two atoms. The electron distribution depends upon all the atoms present in the molecule and the effective electronegativity of a particular atom in other than diatomic molecules will depend on the other atoms to which it is bonded.

The size of atoms

The size of an atom is determined largely by its electronic structure, since most of the space occupied by an atom is taken up by the electron field. Since size depends upon electronic structure, it is a periodic property. There are a number of measures of atomic size, as explained below.

1. Metallic radius

This is half the internuclear distance in the metal. Its value depends to some extent upon the coordination number of the atoms in the metal lattice, i.e. on the way in which the atoms are packed. The values quoted in this text are for coordination number 12. Metallic radii show the following general trends:

(a) They increase down a main group as the number of completed electron shells increase.

(b) They decrease across a period as the nuclear charge increases.

Examples of these trends are illustrated in Table 4.3.

Variations in metallic radii down a group

Li	0·155 nm
Na	0·190 nm
K	0·235 nm
Rb	0·248 nm
Cs	0·267 nm

Variations in metallic radii across Period 4

K	Ca	Ga	Ge
0·235 nm	0·197 nm	0·141 nm	0·137 nm

Table 4.3

2. Ionic radius

The internuclear distance in an ionic crystal may be determined by X-ray diffraction methods. A plot of internuclear distance for various alkali metal ions is shown in Fig. 4.6 The lines corresponding to the various alkali metals are almost parallel. Similarly, the lines corresponding to the various halide ions are almost parallel. This indicates that as the cation changes for the alkali metal halides, there is a constant difference in internuclear distance. This in turn suggests that each ion in the crystal can be assigned a definite radius known as the ionic radius, such that the sum of the ionic radii gives the internuclear distance.

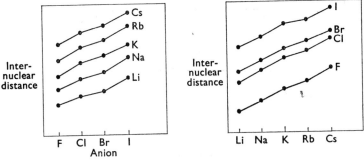

Fig. 4.6

Ionic radii exhibit the following general trends:

(*a*) (i) The cationic radii increase down a main group, as the ionic charge remains constant but the number of completed electron shells increases.

(ii) The cationic radii decrease across a period as the net negative charge on the anion decreases and the nuclear charge increases.

(*b*) (i) The anionic radii increase down a main group, as the ionic charge remains constant and the number of completed electron shells increases.

(ii) The anionic radii decrease across a period as the net negative charge on the anion decreases and the nuclear charge increases.

(*c*) When a metal forms cations of varying ionic charge the ionic radius decreases as the positive charge on the ion increases.

Examples of these various trends are shown in Table 4.4.

Variation of cationic radii for the alkali metals	
Li^+	0·060 nm
Na^+	0·095 nm
K^+	0·133 nm
Rb^+	0·148 nm
Cs^+	0·169 nm
Fr^+	0·176 nm

Variation of anionic radii for the halogens	
F^-	0·136 nm
Cl^-	0·181 nm
Br^-	0·195 nm
I^-	0·216 nm

[*Table continued on page 66*

Variations of cationic radii for Period 5

Rb$^+$	Sr^{2+}	In^{3+}	Sn^{4+}
0·148 nm	0·113 nm	0·081 nm	0·071 nm

Variations of anionic radii for Period 2

N^{3-}	O^{2-}	F$^-$
0·171 nm	0·140 nm	0·136 nm

*Variations of cationic radii for elements showing
variable oxidation number*

	+	2+	3+
Mn		0·080 nm	0·066 nm
Fe		0·076 nm	0·064 nm
Co		0·074 nm	0·063 nm
Ni		0·072 nm	0·062 nm
Cu	0·095 nm	0·069 nm	

Table 4.4

3. Covalent and Van der Waals radii

Fig. 4.7 represents two adjacent diatomic molecules of an element in the crystal lattice of the solid (e.g. solid chlorine). The distance x is the

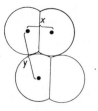

Fig. 4.7

internuclear distance *along the covalent bond axis*. The covalent radius of the atom is equal to $x/2$. The internuclear distance for atoms which are in contact but are not covalently bonded is y. The Van der Waals radius is equal to $y/2$. The covalent radius is always less than the Van der Waals radius. It is roughly additive, as illustrated in Table 4.5.

Molecule	Internuclear distance (nm)	Covalent radius (nm)
Br_2	0·228	0·114
Cl_2	0·199	0·100
I_2	0·267	0·134

Molecule	Calculated bond length $(r_A + r_B)$	Measured bond length
BrCl	$0·114 + 0·100 = 0·214$	0·214
ICl	$0·134 + 0·100 = 0·234$	0·232

Table 4.5

As might be expected, the covalent radius decreases as the number of bonds between the atoms increases. The covalent radius shows the following general trends:

(a) It increases down a main group as the number of completed electron shells in the atoms increases.

(b) It decreases across a period as the increase in nuclear charge decreases atomic size.

The Van der Waals radius shows similar general trends, as illustrated in Table 4.6.

Variations of covalent radius down a group

F	0·072 nm
Cl	0·100 nm
Br	0·114 nm
I	0·113 nm

Variations of Van der Waals radius down a group

F	0·135 nm
Cl	0·180 nm
Br	0·195 nm
I	0·215 nm

Variations of covalent radius across Period 3

Na	Mg	Al	Si	P	S	Cl
0·154 nm	0·130 nm	0·181 nm	0·111 nm	0·106 nm	0·102 nm	0·100 nm

[*Table continued on page 68*

Variations of Van der Waals radius across Period 3

P	S	Cl
0·190 nm	0·185 nm	0·180 nm

Table 4.6

Trends in bond type

Metals and non-metals

The characteristic physical properties of metals are a result of metallic bonding. It was pointed out in Chapter 2 that the two important requirements for metallic bonding were:

(1) relatively small attraction for valency electrons, i.e. low ionization potential,
(2) small number of valency electrons relative to the number of valency orbitals.

This means that metallic character will be a function of the degree of attraction for electrons, i.e. a function of ionization potential or electron affinity or electronegativity.

Ionic and covalent bonds

If sodium is combining with chlorine to give sodium chloride, then, since sodium has a low ionization potential and chlorine a high electron affinity, an ionic compound might be expected. A consideration of electronegativities would lead to the same expectation, since there is a big difference in electronegativity between sodium and chlorine. If chlorine is combining with sulphur, however, then because of the small electronegativity difference, a covalent compound would be expected. In fact, the percentage ionic character which a bond possesses will be directly related to the electronegativity difference between the elements combining. This means that the positions of the combining elements in the Periodic Table will give some indication of the type of bond to be expected, thus:

(1) Elements of groups IM and IIM will combine by means of metallic bonding.
(2) Elements of Groups IM and IIM will combine with elements of Groups VIM and VIIM to give predominantly ionic bonding.
(3) Elements of groups VIM and VIIM will combine to give compounds containing predominantly covalent bonds.

(4) Other elements will combine to give compounds containing bonds which lie on the continuum from predominantly ionic to predominantly covalent, the actual percentage ionic character depending upon the electronegativity difference.

Fajan's rules

Bond type can be predicted in a purely qualitative fashion by the application of Fajan's rules. These state that if two elements combine then ionic bonding is likely if:

(i) the charges on the anion and cation are low,
(ii) the ionic radius of the cation is relatively large,
(iii) the ionic radius of the anion is relatively small.

The rules are based, of course, on the concepts of ionization potential and electron affinity. The likelihood of ionic bonding decreases as the number of charges on the resulting ion increases, because for the positive ion the first, second, third, etc., ionization potentials become progressively larger, and for the negative ion the first, second, third, etc., electron affinities become progressively smaller. Again, if the cation has a small radius this means the tendency to regain the lost electron(s) is increased, and if the anion has a small radius the tendency to eject the gained electrons will decrease (assuming in both cases that distance from the nucleus is the important factor governing attraction for electrons). This amounts to saying that if the radius of the cation is small, then the ionization potential of the original atom will be relatively large, while if the radius of the anion is small, then the electron affinity of the original atom will be large—and this is, broadly speaking, correct.

A fuller consideration of the use of the concepts discussed in this chapter will be given when the general chemistry of the elements is discussed in detail in Chapters 7, 8 and 9.

References

Books

Electronic Structure, Properties and the Periodic Law, H. H. Sisler (Chapman & Hall)
Inorganic Chemistry: An Advanced Textbook, T. Moeller (Wiley)
Advanced Inorganic Chemistry, F. A. Cotton and G. Wilkinson (Wiley)
Introduction to Physical Inorganic Chemistry, K. B. Harvey and G. B. Porter (Addison-Wesley)

Films

Ionization Energy, B. H. Mahan, a CHEM Study Film, available from Sound-Services Ltd., cat. no. 4151/999

5. The solid state

Introduction

Solids are substances which have fixed shape and volume. They possess the properties of relative rigidity and incompressibility, and they cannot flow. The macroscopic properties of solids are thought to result from their underlying structure. The particles (atoms, molecules or ions) making up the solid show little net translational movement relative to each other, but vibrate about fixed mean positions. A time-average picture of a crystalline solid would show that the particles were arranged in a regular pattern. This regular three-dimensional array is called the *crystal lattice*. The arrangement of particles in crystalline solids is said to show long-range order. Certain solids do not possess this property and are therefore non-crystalline, e.g. glass. The under-lying structural order is often reflected in the property of crystal forma-tion on the macroscopic scale. The external form of the crystal thus formed is called the *crystal habit*. The external form (size, area of faces, etc.) is variable, but one constant feature for a given crystal type is the angle between the faces. Some substances seem to show no tendency to form large-scale crystals and for this reason are often said to be *amor-phous*. The lack of obvious large-scale crystalline properties, however, does not preclude the possibility of long-range order on the atomic scale. Amorphous carbon, for example, is thought to be made up of microcrystalline graphite. Crystalline solids are the most important group and will be considered in some detail.

Crystalline solids

General considerations

The bulk properties of crystalline solids will depend upon two general structural factors:

(*a*) the way in which the particles making up the solid are packed together, i.e. the geometrical factor;

(*b*) the type of bonding holding the particles together, i.e. the bonding factor.

The geometrical factor

It will be assumed that the particles making up the crystal are rigid spheres of definite radius. If all the spheres have the same radius, what are the possible ways in which they can be packed together so as to take up as little space as possible? In other words, what are the possible close-packed arrangements for identical spheres?

For one layer of spheres the close-packed arrangement is shown in Fig. 5.1. A given sphere such as X is touching six other spheres in that

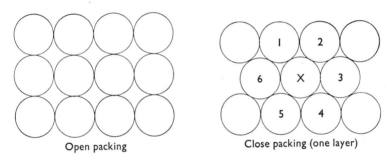

| Open packing | Close packing (one layer) |

Fig. 5.1

layer, i.e. X has six spheres as nearest neighbours. The *coordination number* of X for a single layer of spheres is therefore said to be 6. The second layer of spheres will pack on top of the first as shown in Fig. 5.2.

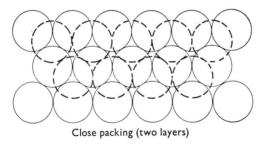

Close packing (two layers)

Fig. 5.2

The third layer of spheres (shaded) can then be placed on the second in one of two ways, as shown in Fig. 5.3.

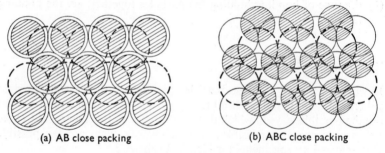

(a) AB close packing   (b) ABC close packing

Fig. 5.3   Two types of close packing

(The shaded spheres are drawn smaller than those in the other two layers for clarity. They are actually touching each other, as in the other layers.)

In (a) the third layer of spheres 'eclipses' the first. If the position of the first layer is denoted by A and that of the second by B, the relative position of the third layer is again A. The position of the fourth layer will be B, of the fifth A, and so on. This type of close packing is called *AB close packing* or *hexagonal close packing*. In (b), the third layer of spheres does not 'eclipse' the first and so the three layers have different relative positions which can be denoted as A, B and C. If the fourth layer then has position A, the fifth position B and the sixth position C, then this is known as *ABC close packing* or *face-centred cubic close*

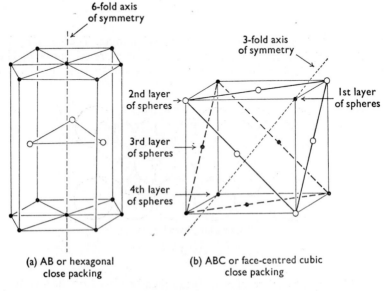

(a) AB or hexagonal close packing

(b) ABC or face-centred cubic close packing

Fig. 5.4

PLATE I
ABAB (or hexagonal) close packing

PLATE II
ABC (or face-centred cubic) close packing

PLATE III
ABC close packing orientated to show the face-centred cubic arrangement

PLATE IV (a)
Sphere in tetrahedral site

PLATE IV (b)
Upper sphere removed to show enclosed sphere

PLATE V (a)
Sphere in octahedral site

PLATE V (b)
Three spheres removed to show enclosed sphere

packing. The nature of AB and ABC close packing may be better understood using a model employing expanded polystyrene spheres or ping-pong balls, but Plates I and II facing p. 72 may make the above discussion clearer. AB and ABC close packing represent two ideal forms, but there is no known reason why the spheres should not be close packed in a less regular way, e.g. ABABABCABABC. However, since most crystal lattices have a regular arrangement, these two ideal types are good starting points for a discussion of lattices in general.

What is the origin of the alternative names for the two types of close packing, viz. 'hexagonal' and 'face-centred cubic'? This becomes clearer if isolated parts of the two types of lattice are examined. Supposing only the centres of the spheres are shown, then the two types of close packing may be represented by Fig. 5.4a and b.

The AB type packing shows hexagonal symmetry; hence the name 'hexagonal' close packing. The spheres in ABC packing are at the corners of a cube with a sphere at the centre of each face; hence the term 'face-centred cubic' close packing. In the latter case the packing planes are at right angles to the body diagonal of the cube. Again, this only becomes obvious when a model is used, though Plates II and III facing p. 72 may help. For both types of close packing 74 per cent of the available space is occupied by the spheres.

Another type of packing which is fairly common is *body-centred cubic packing*. This is not close packing, but is economical of space, in that 68 per cent of the available space is occupied by the spheres. This type of packing is illustrated in Fig. 5.5.

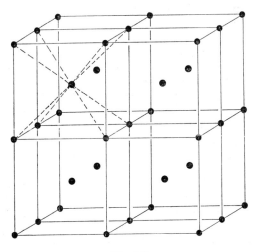

Fig. 5.5

Interstitial sites, coordination number and radius ratio

The usefulness of the three types of ideal packing described above would be limited if their application were confined to lattices made up of identical spheres. Fortunately, by considering the spaces between the spheres, the model can be applied to lattices made up of particles of different radii. An examination of a close-packed array of identical spheres reveals the presence of spaces or 'holes' between the spheres, of various types. These spaces are called *interstitial sites*, and the possible types are illustrated in Fig. 5.6.

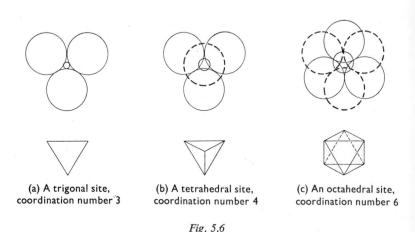

| (a) A trigonal site, coordination number 3 | (b) A tetrahedral site, coordination number 4 | (c) An octahedral site, coordination number 6 |

Fig. 5.6

In a single layer of spheres there is a space between any three touching spheres which could accommodate a fourth sphere, if the radius of the fourth sphere is such that the smaller sphere just touches the other three. This space is called a *trigonal site* and the small sphere would have a coordination number of 3. Again there are sites between two adjacent layers in which a smaller sphere could be placed so that it just touches four larger ones. This type of space is a tetrahedral site and the smaller sphere has a coordination number of 4. There are other spaces between adjacent layers which can accommodate a smaller sphere so that it touches six larger spheres. This type of hole is an octahedral site and the smaller sphere has a coordination number of 6 (see Plates IV (a) and (b) and Plate V (a) and (b), p. 73, for models illustrating tetrahedral and octahedral sites). If the smaller spheres are too large to fit even into the octahedral sites, then there are two further possibilities. Either the close packing must change to a more open form, e.g. with a coordination number of 8, or a 'substitutional compound' rather than an 'interstitial compound' may be formed. In the latter case the smaller sphere simply replaces one of the spheres in the close-packed arrangement. This gives a coordination number of 12. The last two cases are illustrated in Fig. 5.7.

It follows from the above discussion that the coordination number of a given sphere for a lattice made up of more than one type of particle depends upon the relative radii of the spheres. The ratio of the radius of

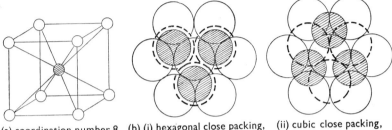

(a) coordination number 8 (b) (i) hexagonal close packing, (ii) cubic close packing, coordination number 12

Fig. 5.7

the smaller particle to that of the larger is called the *radius ratio*. From the solid geometry of the array of spheres, it is possible to calculate the theoretical radius ratio for the occupation of a particular kind of interstitial site or for the formation of a particular kind of packing. These radius ratios and their corresponding coordination numbers are shown in Table 5.1.

Radius ratio	Type of packing	Coordination number
1·00	Close packing	12
0·732	Body-centred cubic	8
0·414	Occupation of octahedral sites	6
0·225	Occupation of tetrahedral sites	4

Table 5.1

The unit cell

A crystal lattice may be described in terms of the smallest part of the lattice out of which it may be built up. This elementary 'building block' is called the *unit cell*. The unit cell for sodium chloride is shown in Fig. 5.8.

Since the empirical formula of sodium chloride is NaCl, the number of sodium and chloride ions in the unit cell must be in the ratio 1:1. This ratio is called the *stoichiometric ratio* for sodium chloride. In general for any unit cell the numbers of the various particles present

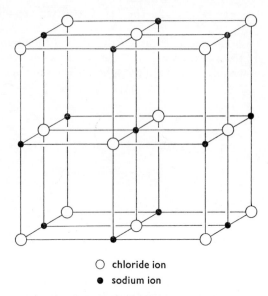

○ chloride ion
● sodium ion

Fig. 5.8   Sodium chloride unit cell

must be in the correct stoichiometric ratio. The fraction of the particle 'belonging' to the unit cell depends upon the particle's position. This is illustrated in Fig. 5.9. Therefore, for sodium chloride,

No. of chloride ions in NaCl unit cell $= 8 \times \frac{1}{8} + 6 \times \frac{1}{2} = 4$
No. of sodium ions in NaCl unit cell $= 12 \times \frac{1}{4} + 1 = 4$

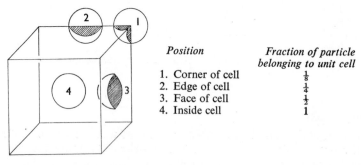

Position	Fraction of particle belonging to unit cell
1. Corner of cell	$\frac{1}{8}$
2. Edge of cell	$\frac{1}{4}$
3. Face of cell	$\frac{1}{2}$
4. Inside cell	1

Fig. 5.9

There are four chloride ions and four sodium ions per unit cell of sodium chloride. This of course gives the correct stoichiometric ratio 1:1.

The number of ions per unit cell and hence the stoichiometric ratio is illustrated for caesium chloride, CsCl, in Fig. 5.10.

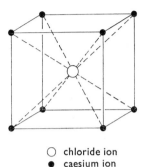

No. of chloride ions = $1 \times 1 = 1$
No. of caesium ions = $8 \times \frac{1}{8} = 1$
∴ Stoichiometric ratio = 1:1

○ chloride ion
● caesium ion

Fig. 5.10

The dimensions of the unit cell may be determined from the result of X-ray diffraction experiments. This gives a means of calculating Avogadro's number, L. Consider the sodium chloride lattice. If l is the distance between the centres of adjacent chloride ions, then

$$\text{volume of the unit cell} = l^3$$

If M is 'molecular' weight of sodium chloride and d is the density of sodium chloride, then

$$\text{volume of 1 mole of NaCl} = Md$$

Therefore

$$\text{number of unit cells/mole} = \frac{Md}{l^3}$$

But, as shown in Fig. 5.8, the number of ion pairs per unit cell is 4, and therefore

$$\text{number of ion pairs per mole} = \frac{Md.4}{l^3}$$

But the number of ion pairs per mole is L, and so

$$L = \frac{Md.4}{l^3}$$

A consideration of the geometrical factor enables one to describe, in a unified and intelligible way, the arrangement of particles in crystal lattices. The lattice arrangement can in turn be used to interpret certain bulk properties such as behaviour under stress, density and crystal habit. However, the interpretation of other properties such as electrical conductivity, melting point and behaviour towards solvents requires the consideration of the second important structural factor, that of bonding.

The bonding factor

Any of the major types of bonding discussed in Chapter 3, metallic, ionic, covalent or Van der Waals, may be responsible for holding the particles in the crystal lattice together. The type of bonding present will give rise in each case to characteristic bulk properties.

1. Metallic crystals

In this case the particles in the lattice are identical, their radii being equal to the metallic radius* of the element concerned. As might be expected, therefore, most metals exhibit either cubic close packing, hexagonal close packing or body-centred cubic packing. This is illustrated in Table 5.2.

Li	Be
bcc	hcp

Na	Mg
bcc	hcp

K	Ca	Sc	Ti	V	Cr	Mn	Fe	Co	Ni	Cu	Zn
bcc	ccp	hcp	hcp	bcc	bcc	—	bcc	hcp	ccp	ccp	hcp

Rb	Sr	Y	Zr	Nb	Mo	Tc	Ru	Rh	Pd	Ag	Cd
bcc	ccp	hcp	hcp	bcc	bcc	—	hcp	hcp	ccp	ccp	hcp

Cs	Ba	La	Hf	Ta	W	Re	Os	Ir	Pt	Au	Hg
bcc	bcc	hcp	hcp	bcc	bcc	hcp	hcp	ccp	bcc	ccp	—

Fr
bcc

Table 5.2   The lattice structures of some of the metals

hcp = hexagonal close packed
ccp = cubic close packed
bcc = body-centred cubic

* The metallic radius is roughly half the internuclear distance in the metal. It depends to some extent on coordination number. The values quoted later in this text refer to coordination number 12.

As indicated in Chapter 3, p. 51, the metal crystal behaves as if it is a regular array of metal ions immersed in a 'sea' or gas of moving valency electrons. This gives rise to the following properties:

(a) Since the bonds are strong but multidirectional, when broken they readily reform. This means that metals are readily extruded into wire and hammered out into sheets, i.e. they are ductile and malleable.

(b) The mobile electrons in the 'electron gas' will move readily if a potential difference is applied across a metal; i.e. metals are good conductors of electricity.

(c) The 'electron gas' will rapidly transmit thermal energy, so that metals are good conductors of heat.

(d) The metallic bonds are strong, so that the melting points and boiling points of metals are relatively high.

(e) Because the atoms are close packed, or nearly so, the densities of metals are relatively high.

(f) The mobile valency electrons are easily excited to higher energy levels by incident radiation, which is therefore absorbed. This means that metals will be opaque. The excited electrons, on returning to lower energy levels, then emit radiation, so giving rise to the lustre characteristic of metals.

2. Ionic crystals

Ideally, these are crystals having lattices made up of positive and negative ions. In general, the ionic radii of the two ions will be different, so that the type of packing and hence the coordination number of the ions will depend upon the radius ratio. The particles are charged so that two conditions must be fulfilled; (a) the number of ions of each kind must be such that the total charge is zero, and (b) the positive ions tend to be surrounded by negative ions as nearest neighbours, and vice versa. Some examples of ionic crystals are discussed below.

Sodium chloride, NaCl

The structure of sodium chloride is illustrated in Fig. 5.11. Chloride ion A has six sodium ions as its nearest neighbours, i.e. it has a coordination number of 6. Similarly, sodium ion B has six chloride ions as its nearest neighbours and so also has a coordination number of 6. If the lattice were continued indefinitely every sodium ion would be found to be surrounded by six sodium ions and vice versa. The ionic radius of sodium is 0·095 nm and that of chlorine 0·181 nm. Therefore the radius ratio is 0·095/0·181, i.e. 0·55. Reference to Table 5.1 indicates that ideally the radius ratio should be 0·414 for occupation of octahedral sites. 0·55 is reasonably close to this, so that the sodium chloride lattice may be considered to be an array of cubic close-packed chloride ions with the octahedral sites occupied by sodium ions, giving a coordination number of 6. Looked at in another way, the NaCl lattice may be thought of as two inter-penetrating lattices, one made up of cubic close-packed sodium ions and the other of cubic close-packed chloride ions. The unit

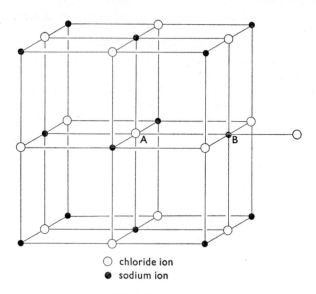

○ chloride ion
● sodium ion

Fig. 5.11   The sodium chloride lattice

cell of NaCl has already been considered and gives the correct stoichiometric ratio 1:1 (p. 76).

Caesium chloride, CsCl

Although caesium is an alkali metal like sodium, its chloride does not have the sodium chloride structure. This is accounted for by the much bigger ionic radius of caesium, 0·169 nm compared with 0·095 nm for sodium. This gives a radius ratio of 0·169/0·181, i.e. 0·93 approximately. This is well outside the limits for a sodium chloride type structure, so caesium chloride adopts a more open packing as shown in Fig. 5.10. The stoichiometric ratio is shown to be 1:1 on p. 77.

Zinc sulphide, ZnS

There are two different forms of zinc sulphide, zinc blende and wurtzite; these have different structures.

(i) *Zinc blende*   This illustrates a third type of structure for 1:1 stoichiometry. The sulphur atoms are cubic close packed. Each atom in a close-packed structure has two tetrahedral sites associated with it.

(This may be seen by examining a model of close-packed spheres.) Half of the available tetrahedral sites are occupied by zinc atoms, thus giving a 1:1 stoichiometry.* This is illustrated in Fig. 5.12.

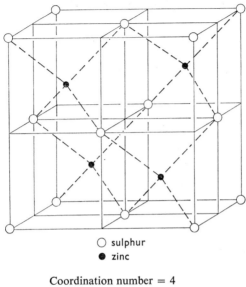

○ sulphur
● zinc

Coordination number = 4
Stoichiometry:
8 'corner' sulphurs = $8 \times \frac{1}{8}$ = 1
6 'face' sulphurs = $6 \times \frac{1}{2}$ = 3
_____

total sulphur = 4

4 'inside' zincs = 4
∴ Stoichiometry = 4:4 = 1:1

Fig. 5.12   The zinc blende structure

(ii) *Wurtzite*   This is similar to zinc blende. In this case the sulphur atoms are hexagonal close packed instead of cubic close packed. The zinc atoms again occupy half the available tetrahedral sites.

Fluorite, CaF₂

Here, the stoichiometry is 1:2. The calcium ions adopt the cubic close-

* The radius ratio (Table 5.1) for occupation of tetrahedral sites should be of the order 0·225. The radius ratio for ZnS is much larger than this. However, the predominant factor here is bonding. The zinc atom bonds by means of sp^3 hybrid orbitals which have a tetrahedral configuration. (See zinc complexes, Chapter 9, p. 180.)

packed arrangement and the fluoride ions occupy *all* the tetrahedral sites, thus giving the correct stoichiometric ratio.

Properties of ionic compounds

Ionic compounds are made up of ions attracted together by powerful electrostatic forces. The lattice arrangement depends upon the factors discussed above. They will possess a number of characteristic properties:

(*a*) They possess definite lattice arrangements, and therefore each compound occurs in one or more definite crystalline forms.
(*b*) They possess no mobile electrons, and so are poor conductors of electricity, in the solid state.
(*c*) The bonding forces are strong, so they tend to have high melting and boiling points.
(*d*) If the ions are made mobile, they can act as current carriers and the resulting system would be a good conductor of electricity. The ions may become mobile either by fusion of the solid or by solution in a suitable solvent.
(*e*) Ionic solids show little tendency to dissolve in covalently bonded liquids, but in polar solvents such as water, which interact strongly with the constituent ions, they often dissolve quite readily.

3. Covalent crystals

In some crystals the bonds joining the constituent atoms together are covalent. A typical example of this is diamond, which is made up of a regular array of carbon atoms, each carbon being bonded to four others by single covalent bonds. The structure is illustrated in Fig. 5.13.

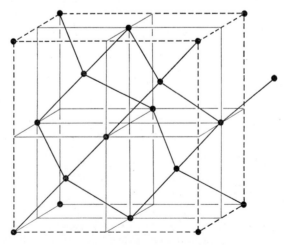

Fig. 5.13   Diamond structure

There are no discrete molecules present, all the atoms being bound together by a network of powerful covalent bonds. Such a structure is often termed a *macromolecular* or *giant-molecule* structure.

Another example is the other allotrope of carbon, graphite. This allotrope is made up of macromolecular sheets of carbon atoms with much weaker bonding between the sheets. This is illustrated in Fig. 5.14.

Each carbon only uses three of its valency electrons in this structure. The fourth electron from each carbon is delocalized across the whole sheet. The properties of diamond and graphite are contrasted and correlated with structure on p. 130, Chapter 8.

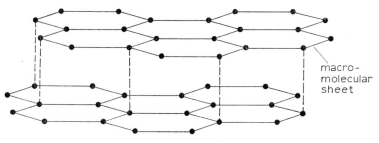

macro-molecular sheet

Fig. 5.14   Graphite

Properties of covalent crystals

In a completely covalent crystal, such as diamond, the atoms are bound together by strong covalent bonds and all the valency electrons are highly localized as bonding pairs. Such compounds therefore possess the following characteristic properties:

(a) Because of the strength of the bonds they are extremely hard and have very high melting points and boiling points and relatively high densities. They are chemically inert.

(b) Since there are no mobile electrons they are poor conductors of heat and of electricity. Incident light is not absorbed and so they are often transparent.

4. Molecular crystals

Many substances are made up of discrete molecules. The atoms in the molecule are bound together by strong intra-molecular covalent bonds. Most non-metallic elements and a large number of organic compounds are of this type. When these substances form a solid, the molecules are attracted together by relatively weak Van der Waals forces. Such substances form molecular crystals. Iodine is a typical example, and its structure is illustrated in Fig. 5.15.

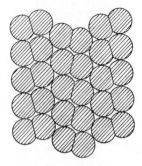

Fig. 5.15   Layer of iodine molecules in crystal of iodine

Properties of molecular crystals

The intermolecular forces are weak and there are no mobile electrons. Molecular solids possess the following characteristic properties:

(*a*) They have relatively low densities and low melting and boiling points. They sometimes sublime on heating.
(*b*) The constituent molecules do not interact strongly with polar solvents such as water. They are often insoluble or sparingly soluble in water, but soluble in organic liquids.
(*c*) They are relatively soft.
(*d*) They are poor conductors of electricity in solution and when molten.

The band model of solids

A quantitative treatment of the 'electron gas' theory of metallic structure gives incorrect results for the specific heats of metals. The application of quantum mechanics resolves this difficulty and gives a more sophisticated model of the structure of solids in general. For isolated atoms, each of the electrons is in a discrete energy state or level. For an aggregate of atoms, quantum mechanics indicates that a relatively broad energy band corresponds to each of the narrow energy levels in the isolated atom. This is represented diagrammatically in Fig. 5.16.

These energy bands are occupied by all the electrons present in the

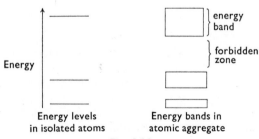

Fig. 5.16

metal, the lower levels being occupied first. This model explains the different electrical properties of solids. In a metal such as sodium, the original valency orbitals are only half-full in the isolated atoms. In the metal, therefore, the uppermost energy band is only partly filled. This means that if an electric field is applied, the electrons are able to move into higher energy states within the uppermost band and so an electric current flows. Sodium is a good conductor of electricity. For a substance like diamond on the other hand, all the energy bands are completely full. The next empty band is separated from the uppermost filled band by a large forbidden zone. There is no flow of electrons when an electric field is applied to diamond; thus diamond is an insulator.

The electrical conductivity of metals decreases with increase in temperature. This is because thermal agitation of the atoms scatters the conducting electrons. However, the electrical conductivity of certain substances increases with increase in temperature. These substances are known as *semiconductors*. This behaviour can be explained by the band model, if it is assumed that semiconductors are substances with filled energy bands but with a relatively small forbidden zone between the uppermost filled band and the next empty band. Thermal energy enables some of the electrons in the uppermost level to jump the forbidden zone, if this is not too large, and enter the next empty band. The promoted electrons will leave electron 'holes' behind them. The two outer bands are now only partly filled and so the substance will conduct. The number of promoted electrons increases with increase in temperature and so the conductivity increases.

The nature of conductors, insulators and semiconductors is represented diagrammatically in Fig. 5.17.

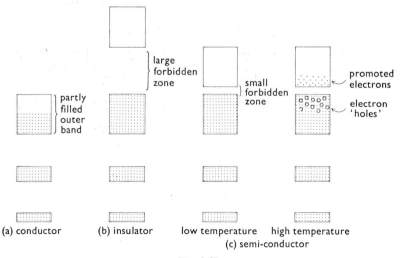

Fig. 5.17

Lattice defects

In the discussion of solids above it was assumed that all the lattices were ideal, and a purely static treatment was given. At normal temperatures, however, the particles present are vibrating about mean positions. The lattice as a whole is in a state of relatively violent agitation. The chances are that some of the particles will be displaced and the lattice will develop irregularities. Such irregularities are called *lattice defects*.

Stoichiometric defects

These are defects occurring without any change in the stoichiometry of the crystal. These are of two main types, *Schottky defects* and *Frenkel defects*. They are illustrated in Fig. 5.18.

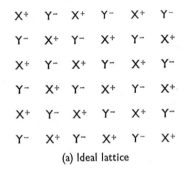

(a) Ideal lattice

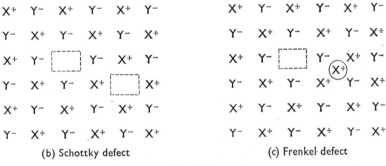

(b) Schottky defect   (c) Frenkel defect

Fig. 5.18

A Schottky defect occurs when a pair of ions of opposite charge is 'missing' from the lattice. A Frenkel defect occurs when a cation occupies one interstitial site and so leaves a gap in the lattice. Frenkel defects are most common in substances where the cation is relatively small and so can readily occupy an interstitial position, and where the coordination number is low, so that the electrostatic attraction to be overcome for movement to the interstitial position is relatively low. Zinc sulphide, for example, with a small cation and a coordination number of 4, shows Frenkel defects. On the other hand, sodium chloride, with a coordination number of 6 and ions of more similar size, shows Schottky defects. The number of defects is very small at room temperature but rises quite rapidly as the temperature increases, being about one per 10 000 lattice sites for sodium chloride at 800°C.

Non-stoichiometric defects

When these occur the expected stoichiometric ratio for the compound is changed. The resulting compounds are called non-stoichiometric, non-Daltonian or Berthollide compounds, and they contravene the laws of chemical combination as enunciated by Dalton. Iron(II) oxide at normal temperature and pressure has a composition which varies between $Fe_{0.94}O$ to $Fe_{0.84}O$, whereas the theoretical composition should be FeO. Variations from the theoretical stoichiometric ratio can be explained in terms of lattice defects. These are of four main types, illustrated in Fig. 5.19.

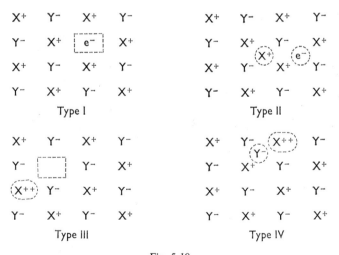

Fig. 5.19

Type I has an excess of metal due to anion vacancies. Electrons remain in the vacated anion sites so that the electrical neutrality of the lattice is maintained. Type II has a metal excess due to the presence of interstitial cations, together with the appropriate number of associated electrons to preserve neutrality. In type III there is metal deficiency due to cation vacancies. Electrical neutrality is preserved by an increase in the oxidation state of other cations present. Metal deficiency is caused in type IV by the presence of interstitial anions. Again, increase in the oxidation state of some of the cations present preserves electrical neutrality.

Compounds showing type I metal excess are not very common. Non-stoichiometric sodium chloride of this type may be prepared, however, by treating sodium chloride with sodium vapour. Compounds which might be expected to show Frenkel defects often exhibit non-stoichiometry of type II, e.g. ZnO. The free electrons in compounds of types I and II may absorb incident radiation, so that the substance will be coloured. The peculiar behaviour of zinc oxide, which changes from white to yellow when heated, and back to white on cooling, is probably to be accounted for by this phenomenon. Type III requires that the cation have variable valency and might therefore be expected with transition metal compounds. This is indeed the case. Examples of compounds exhibiting this type of defect are iron(II) oxide and nickel(II) oxide. Type IV also requires the cation to exhibit variable valency. There are no known examples of crystals showing this type of defect, however. This is probably because the anion is usually larger than the cation and so does not easily move into an interstitial position.

Non-stoichiometric compounds belong to the class of substances known as semiconductors. Substances with metal excess defects contain free electrons, which give rise to the conducting properties. Since the mechanism of conduction is similar to normal conduction by a metal, they are called normal-type or n-type semiconductors. In the metal-deficient substances there are no free electrons, so that conduction occurs by transfer of an electron from a cation with the lower oxidation number to one with the higher. A positive charge, therefore, moves in the opposite direction to the electron. This is known as positive hole conduction. Such substances are known as positive-hole or p-type semiconductors.

References

Books

Structural Principles in Inorganic Compounds, W. E. Addison (Longmans)

Chemistry Today, Organisation for Economic Co-operation and Development

An Introduction to Modern Chemistry, M. J. S. Dewar (Athlone Press)

Papers

'The nature of solids', G. H. Wrunier (reprint from *Scientific American*, December 1952)

Films

Crystals and Their Structures, J. A. Campbell, CHEM Study Film. Available from Sound-Services Ltd., cat. no. 4139/999

Film strips

Atomic and Molecular Models, a series of seven film strips from Encyclopaedia Britannica Films

6. The nucleus

Composition of the nucleus

In the discussion of atomic structure in Chapter 1, it was pointed out that the diameter of the central part of the atom, the nucleus, is very much smaller than the overall diameter of the atom itself. Investigation of the scattering effect of the nucleus on bombarding particles shows that the nuclear diameter is of the order of 10^{-15}m, compared to an average over-all atomic diameter of 10^{-10}m. The nucleus is a composite structure made up of so-called fundamental particles. The simpler aspects of nuclear chemistry can be described in terms of four of these (excluding electromagnetic radiation which is made up of photons), though thirty-three such particles have so far been identified. The properties of these four particles are listed in Table 6.1.

Name	Mass (relative to ^{12}C)	Charge (relative to +1 charge on proton)
Proton	1·007825	+1
Neutron	1·008665	0
Electron	very small	−1
Positron	very small	+1

Table 6.1

The term *nucleon* is used as a general name for any particle present in the nucleus.

If the number of neutrons (N) in the nucleus, is plotted against the number of protons (P) in the nucleus, a graph of the form shown in Fig. 6.1 is obtained.

For the lighter elements the neutron/proton ratio is one, or nearly so, for stable nuclei. However, as the mass number increases there is an increasing preponderance of neutrons over protons. This trend can be explained, in part, by a consideration of intra-nuclear forces. Nothing is known of the *origin* of the attractive forces operating in the nucleus but it is known that (*a*) they are *not* electrical or gravitational in origin, (*b*) they are extremely powerful and (*c*) they are very short range. Suppose a proton approaches an atomic nucleus from a relatively great

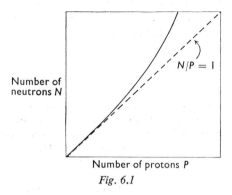

Number of neutrons N

Number of protons P

Fig. 6.1

distance, then at first the two positively charged particles exert a repulsive force upon each other and the potential energy of the proton increases. However, when the distance between the particles approximates to the nuclear diameter, namely 10^{-15}m, the powerful attractive nuclear forces predominate and the proton is attracted into the nucleus. This process is illustrated in Fig. 6.2a. The process for the approach of a neutron is similar, but since the neutron has zero charge there is no electrostatic repulsive force to overcome. This is illustrated in Fig. 6.2b.

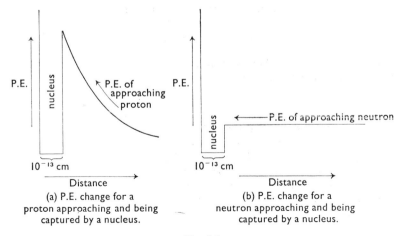

(a) P.E. change for a proton approaching and being captured by a nucleus.

(b) P.E. change for a neutron approaching and being captured by a nucleus.

Fig. 6.2

The electrostatic repulsive forces between protons do not disappear when the proton is within the nucleus, but are very much less than the attractive nuclear force. This goes some way towards explaining why the ratio N/P is unity for the lighter elements, but increases for the heavier elements. As the number of protons increases, so the repulsive forces between them increase, tending to make the nucleus unstable. This effect is counterbalanced by the presence of excess neutrons with consequent increase in attractive forces and no increase in associated repulsive forces.

Binding energy

It is found that the mass of a nucleus is not equal to the sum of the masses of the particles making up the nucleus. Consider, for example, ^2H. This is deuterium, an isotope of hydrogen, with a nucleus made up of one proton and one neutron. Reference to Table 6.1 predicts that the nuclear mass should be given thus:

| mass of 1 proton | = 1·007825 a.m.u. (atomic mass units) |
| mass of 1 neutron | = 1·008665 a.m.u. |

∴ predicted total mass
  of nucleus = 2·016490 a.m.u.

Mass spectroscopy gives an experimental value for the nuclear mass of deuterium of 2·014102 atomic mass units, which is less than the calculated mass. The *mass defect* may be calculated by subtracting the experimental value from the calculated one, thus:

calculated mass = 2·016490 a.m.u.
experimental mass = 2·014102 a.m.u.

mass defect = 0·002388 a.m.u.

Einstein has shown that there is an equivalence between mass and energy given by the equation

$$E = mc^2$$

where E = energy, m = mass and c = the velocity of light. The loss in mass represented by the mass defect represents the energy binding the nucleus together. This is known as the *binding energy*. The unit of energy usually used in nuclear chemistry is the electron volt (eV). This is defined as the energy change undergone by an electron when it is accelerated across a potential difference of one volt. Using the Einstein equation it can be shown that:

1 a.m.u. = 931 MeV (million electron volts)

The binding energy for deuterium therefore is given by:

$$\text{binding energy} = 0\cdot0023 \times 931 \text{ MeV}$$
$$= 2\cdot2 \text{ MeV (approx.)}$$

Then, since 1 eV $=96\cdot48$ kJ mol^{-1}:

$$\text{binding energy} = 2\cdot2 \times 10^6 \times 96\cdot48$$
$$= 212\cdot3 \times 10^6 \text{ kJ mol}^{-1}$$

Compare this with the bond energy for chemical bonds, which is from 200 to 650 kJ mol^{-1} approximately. The nuclear binding energy is a million times as great as the average chemical bond energy.

It is reasonable to assume that as the number of nucleons in the nucleus increases, so the total binding energy will increase. This is indeed so, but a graph of binding energy *per nucleon* against mass number has the form indicated in Fig. 6.3. For the lighter elements the bind-

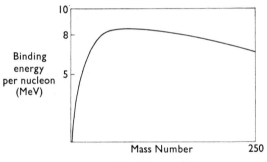

Fig. 6.3

ing energy per nucleon increases rapidly, but after mass number 25 the curve levels off and then falls gradually for the heavy elements. The graph indicates that the fusion of nuclei of the lighter elements to give heavier nuclei will occur with the evolution of energy. It is said to be an exoergic process. The fission of heavy nuclei to give lighter ones will also often be exoergic. Elements having mass number lying on the flat part of the curve will, on the other hand, undergo both fission and fusion processes with the absorption of energy. These processes are said to be endoergic.

Models of nuclear structure

The liquid drop model

Fig. 6.3 shows, that for all but the lightest and heaviest elements, the binding energy per nucleon is approximately constant and equal to

8 MeV. This, combined with the fact that the density of nuclear matter is approximately constant, suggests a possible analogy between the behaviour of nucleons in the nucleus and the behaviour of molecules in a liquid drop. Since the binding energy per nucleon is roughly constant then the total binding energy is proportional to the number of nucleons present. For a liquid the total heat of vaporization, which corresponds to the binding energy, is proportional to the volume of the liquid, i.e. to the number of molecules present. The model therefore suggests that the nucleus behaves as a drop of nuclear fluid, the 'molecules' of which are the nucleons. Since the drop is very small, 'surface tension' will be an important factor. This model can be used to interpret the shape of the curve in Fig. 6.3. The binding energy per nucleon rises rapidly for the lighter elements as the surface tension factor becomes increasingly important. Eventually it reaches the saturation value of 8 MeV and then begins to fall for the heavier elements as the electrostatic forces of repulsion between the nucleons become more important. The phenomenon of nuclear fission can be interpreted in terms of the oscillations of the liquid drop, the restoring forces being due to 'surface tension'.

The shell model

The observed angular momenta and associated magnetic properties of nuclei indicate that the liquid drop model does not give a complete description of nuclear properties. To interpet these phenomena it is necessary to use a shell model, similar in many ways to the shell model of electron structure. Just as there are relatively very stable electronic structures for atoms, e.g. noble gas structures, so there are relatively very stable nuclear structures. The nuclear shells are closed at the so-called magic numbers:

N (number of neutrons) or Z (number of protons)
$$= 2, 8, 20, 28, 50, 82 \text{ or } 126$$

For example, one of the isotopes of oxygen is $^{16}_{8}O$, i.e. N and $Z=8$, so this is a very stable nucleus, and consequently $^{16}_{8}O$ is one of the most abundant nuclei. Again for $^{208}_{82}Pb$, $N=126$ and $Z=82$ so that this isotope lies at the end of a radioactive series.

A combination of the liquid drop and shell models can be used to describe nuclear reactions. For example, suppose aluminium is bombarded with neutrons, then there are a number of possible reactions:

$$^{27}_{13}Al + ^{1}_{0}n \longrightarrow [^{28}_{13}Al]^* \text{ excited state}$$

$$\longrightarrow ^{28}_{13}Al + \gamma \qquad (1)$$

$$\longrightarrow ^{27}_{12}Mg + ^{1}_{1}p \qquad (2)$$

$$\longrightarrow ^{24}_{11}Na + ^{4}_{2}\alpha \qquad (3)$$

$$\longrightarrow ^{26}_{13}Al + 2^{1}_{0}n \qquad (4)$$

where γ = radiation, charge 0, mass 0
$\quad {}_1^1p$ = a proton, charge 1, mass 1
$\quad {}_2^4\alpha$ = an α particle, charge 2, mass 4
$\quad {}_0^1n$ = a neutron, charge 0, mass 1.

The other symbols have the usual meaning. The first change may be represented diagramatically as in Fig. 6.4.

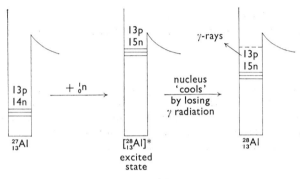

Fig. 6.4

The other possible reactions 2, 3 and 4 all involve the loss of nucleons. The excited nucleus decreases its energy by the 'evaporation' of a nucleon; i.e. one or more particles may obtain sufficient energy to overcome the 'surface tension' and so escape from the nucleus, leaving it with less energy.

Radioactivity

Many isotopes are unstable and they eventually disintegrate to give more stable products. When this occurs spontaneously the phenomenon is known as natural radioactivity. Artificial radioactive isotopes may be prepared by bombarding otherwise stable isotopes with high-energy particles, either in a particle accelerator or in a nuclear reactor. Radioactive processes may be divided into four main types.

(a) α-particle emission

The nucleus ejects an α-particle, giving a product with a mass number four units less than the starting element and an atomic number two units less; e.g.

$${}_{88}^{226}\text{Ra} \longrightarrow {}_{86}^{222}\text{Rn} + {}_2^4\alpha$$

(b) β-particle (electron) emission

The nucleus, of course, contains no electrons, but neutrons themselves may undergo the change

$$^{1}_{0}n \longrightarrow ^{1}_{1}p + ^{0}_{-1}e$$

If the proton remains in the nucleus this amounts to the ejection of an electron; e.g.

$$^{32}_{15}P \longrightarrow ^{32}_{16}S + ^{0}_{-1}e$$

(c) Positron emission

Again, the nucleus does not contain positrons as such, but protons may undergo the change

$$^{1}_{1}p \longrightarrow ^{1}_{0}n + ^{0}_{1}e$$

This process does not occur for free protons since it is endoergic (energy absorbing), but it may well occur in a complex nucleus. Here, the binding energy of the nucleus may increase sufficiently, by a decrease in the electrostatic repulsive forces due to the ejection of a proton, to make the overall process energetically favourable; e.g.

$$^{18}_{9}F \longrightarrow ^{18}_{8}O + ^{0}_{1}e$$

There are no naturally occurring positron emitters.

(d) K-capture or electron capture

The nucleus may capture an electron from the innermost or K shell of the electron field; e.g.

$$^{7}_{4}Be + ^{0}_{-1}e \longrightarrow ^{7}_{3}B$$

All the above processes occur because the unstable nucleus changes, in order to produce a better balance between the attractive nuclear forces between nucleons in general, and the electrostatic repulsive forces between protons.

Applications of nuclear chemistry

Chemical applications

Radio-isotopes are useful because they can be detected when present in very small concentrations, through their radioactivity. This gives rise to many important applications.

(a) Study of equilibria

If barium sulphate is allowed to come to equilibrium with its saturated solution, then according to chemical theory, although there is no *net* change in the concentrations of barium and sulphate ions in solution, yet, because the equilibrium is a dynamic one, there is still a continual interchange of barium and sulphate ions between the solid and solution phases. This can be shown by introducing 'labelled' barium sulphate, i.e. barium sulphate made from a radio-isotope of barium, into the system. If a quantity of *the solution* is examined after some time has elapsed, it is found to be radioactive. This indicates that interchange of barium ions has occurred.

The vapour pressures of metals such as iron can be studied at quite low temperatures if the metal is labelled with a radio-isotope. Again, equilibria involving the absorption of ions or molecules on solid surfaces can be studied using radioactive measurements.

(b) Study of reaction mechanisms

If heavy water is mixed with ethyl alcohol and the mixture then chemically separated, it is found that the ratio of ordinary hydrogen to deuterium (heavy hydrogen) in the water is approximately equal to the ratio in the hydroxyl groups of the alcohol. There is no deuterium present in the ethyl group of the alcohol. This shows that exchange of hydrogen ions between water and the hydroxyl groups takes place rapidly, but no such exchange occurs between water and the ethyl groups. In a similar way the labelling of particular atoms helps in deciding probable reaction mechanisms in many instances.

(c) Quantitative analysis

One important analytical technique using radio-isotopes is *activation analysis*. The principle here is to irradiate a sample containing an unknown quantity of the substance under consideration. The bombarding particles are often slow neutrons. This process converts the substance into a radioactive isotope. If the activity of the sample is then measured and compared with the activity of a previously irradiated control sample containing a known quantity of the substance, then the proportion of the substance in the unknown can be calculated. The proportion of manganese in a relatively pure sample of aluminium can be calculated by this method. The aluminium is irradiated with neutrons, when the manganese is converted into a radio-isotope, thus:

$$\,_{25}^{55}\text{Mn} + \,_{0}^{1}\text{n} \longrightarrow \,_{25}^{56}\text{Mn}$$

The radio-manganese emits β-rays and has a half-life of 2·58 hours. The aluminium also becomes radioactive, but the radio-isotope in this case has a half-life of only 2–3 minutes. Therefore, if the sample is left for some time before measurement of activity, the proportion of manganese can be estimated without destruction of the sample.

Another important analytical technique is the *isotope dilution* method. The following example illustrates this method. It is very difficult to analyse mixtures of the rare earths, neodymium and praseodymium quantitatively by chemical techniques. However, it is possible to isolate pure praseodymium from the mixture by chemical methods, if a quantitative yield is not required. The original mixture is treated with a small quantity of radioactive praseodymium. A quantity of praseodymium is now isolated from the mixture by ordinary chemical methods and weighed. This will contain the appropriate proportion of radio-praseodymium, since the radio-isotope behaves chemically in the same way as the non-radioactive praseodymium. If the activity of the isolated praseodymium is then measured, it follows that:

$$\frac{\text{weight of praseodymium in original sample}}{\text{weight of isolated praseodymium}}$$

$$= \frac{\text{activity of original radio-isotope added}}{\text{activity of isolated sample (corrected for decay if necessary)}}$$

Hence the weight of praseodymium in the original sample may be found.

An interesting application of this method is the determination of blood volume in man. A quantity of saline solution containing a known quantity of ^{55}Cr-labelled blood cells is injected into the patient. Sufficient time is then allowed to elapse for the labelled cells to mix uniformly with the unlabelled ones. A known volume of blood is then withdrawn and the activity measured. Then,

$$\frac{\text{total volume of blood}}{\text{volume withdrawn}} = \frac{\text{initial activity of injected saline}}{\text{activity of sample withdrawn}}$$

Radio-isotopes are very important in determining the rate of distribution of various substances in the living animal or human body. If an animal is fed with labelled calcium then the rate at which this appears in the various bones can be followed. The rate at which water reaches the various parts of the body has been determined by letting the animal drink water labelled with tritium ($^{3}_{1}H$). Calvin has investigated the primary products in plant photosynthesis using radioactive carbon dioxide.

Technological applications

The applications in this field are numerous, but two examples must suffice. It is important to the engineer to know the rate of wear of the various moving parts in an engine. Suppose the rate of wear in the bearings is required. If the bearings are labelled by adding a small quantity of radioactive metal in their manufacture, then the rate of wear can be measured by measuring the increase in activity of the lubricating oil due to the presence of metal dust from the bearings.

Another interesting application is the radioactive thickness gauge. This enables the variation in thickness of, say, a moving steel sheet to be followed. A γ source is placed on one side of the sheet and a radiation detector on the other. The radiation intensity at the detector depends upon the thickness of the sheet between source and detector. The detector can be calibrated so that the thickness can be read off directly. Since the measurement of thickness is continuous, the rollers producing the sheet can be adjusted accordingly to give a sheet of uniform thickness.

Radio-carbon dating

The atmosphere is continually being bombarded with cosmic rays from outer space. These interact with the atmosphere to produce neutrons. The neutrons interact with atmospheric nitrogen to produce the radio-isotope $^{14}_{6}C$:

$$^{14}_{7}N + ^{1}_{0}n \longrightarrow ^{14}_{6}C + ^{1}_{1}p$$

The $^{14}_{6}C$ isotope then decays to give back nitrogen:

$$^{14}_{6}C \longrightarrow ^{14}_{7}N + ^{0}_{-1}e$$

These two processes are occurring continuously, and the rate of formation equals the rate of decay, so that the concentration of $^{14}_{6}C$ in the atmosphere is constant. The half-life (time for a given initial concentration to decrease to one half of its original value) for the decay of $^{14}_{6}C$ is about 5700 years. The radio-carbon is assimilated by plant and animal life, all of which contain a fixed proportion of this isotope. If the organism dies no fresh carbon is assimilated and so the $^{14}_{6}C$ present decays without being replaced. Carbon from an organism which died 5700 years ago will be only half as radioactive as carbon from a living organism. Thus, measurement of the radioactivity of the carbon enables the time which has elapsed since the organism died to be calculated.

Fission and fusion processes

The most spectacular examples of nuclear reactions are those involved in the atomic and hydrogen bombs. The atomic bomb employs a *fission* process. An isotope of uranium ^{235}U reacts with neutrons to split into two fragments and release more neutrons. The neutrons so released can react with other ^{235}U nuclei producing further fission. Since uranium is one of the heaviest elements this process is exoergic (energy producing). If the piece of ^{235}U is above a certain critical size (less than one kilogramme), the process becomes self-propagating and a chain reaction ensues. In the fission bomb portions of ^{235}U less than the critical amount are kept apart from each other until such time as a device is triggered to bring them together. The ^{235}U is then larger than the critical size, so that a rapid chain reaction and subsequent explosion occurs.

The thermal energy of the sun is produced by a fusion process:

$$4\,^{1}H \longrightarrow \,^{4}He$$

A considerable loss in mass occurs when four hydrogen nuclei are converted into one helium nucleus, so that the process is highly exoergic. This process occurs only at very high temperatures. In the hydrogen bomb a fusion process is used. It is thought to be

$$^2H + {}^6Li \longrightarrow 2\,{}^4He$$

This process requires a temperature of about $100\,000\,000°C$ for it to occur explosively. A fission bomb is used as a 'fuse' to attain the required ignition temperature.

In nuclear reactors fission reactions are used. One possible nuclear fuel is uranium, of the natural isotopic composition, mixed with a moderator such as graphite, which enables the chain reaction to develop fairly slowly. The reaction can be controlled by inserting rods of neutron-absorbing materials such as boron or cadmium. A coolant, such as gaseous carbon dioxide, is circulated around the reactor, and the hot gas is used to produce steam which in turn drives turbines, so producing electricity.

References

Books

Basic Concepts of Nuclear Chemistry, R. T. Overman (Chapman and Hall)

Chemistry Today, Organisation for Economic Co-operation and Development

An Introduction to Modern Chemistry, M. J. S. Dewar (Athlone Press)

Concise Inorganic Chemistry, J. D. Lee (Van Nostrand)

College Chemistry, L. Pauling (Freeman)

Papers

'The structure of the nucleus', B. H. Flowers, from *Collected Readings in Inorganic Chemistry*, p. 201 (American Chemical Society)

7. Hydrogen and the s-block elements

Introduction

The aim of this chapter and of Chapters 8 and 9 is to survey the general chemistry of the better known elements. An attempt will be made to correlate the electronic characteristics of the elements with their properties. Each section is divided into three parts. First of all a table is given which summarizes some of the fundamental numerical characteristics. Then follows a discussion of important structural points, and finally a survey of the chemistry of the group being considered.

1. Hydrogen

Symbol	Electronic structure	First ionization energy (kJ mol^{-1})	Electro-negativity (Pauling)	RADII (nm)		
				Ionic (−1)	Covalent	Van der Waals
H	1s^1	1310	2·1	0·208	0·037	—

Table 7.1

Structural characteristics

(i) A hydrogen atom has one electron in its valence shell. There are no underlying electron shells. If the valence electron is removed a proton is left.

(ii) The first ionization energy is relatively high, much higher than the maximum usually associated with metals (1050 kJ mol^{-1}).

(iii) Hydrogen has a very small covalent radius and is, indeed, the smallest of all atoms.

(iv) The radius of the hydride ion (H$^-$) lies between that of bromide, Br$^-$ (0·195 nm), and iodide, I$^-$ (0·216 nm).

(v) The valence shell of hydrogen is complete when it contains two electrons.

(vi) Hydrogen is moderately electronegative (H, 2·1; Cs, 0·7; F, 4·0 on the Pauling scale).

General chemistry of hydrogen

Hydrogen can attain a noble gas structure in two possible ways: (1) by sharing an electron pair with another atom, and (2) by gaining an electron to form the hydride ion, H^-. In each case the valence shell of hydrogen has a helium structure, that is, $1s^2$. The former is the more common way in which hydrogen bonds with other atoms. Because of the moderate value of its electronegativity hydrogen can form stable covalent bonds with elements of moderate electronegativity and those of relatively high electronegativity; e.g. with carbon in methane,

$$\begin{array}{c} H \\ | \\ H-C-H \\ | \\ H \end{array}$$

and with chlorine in hydrogen chloride gas,

$$\overset{\delta+}{H}\!\!-\!\!-\!\!-\!\!-\!\!\overset{\delta-}{Cl}$$

However, in the latter case, because of the difference in electronegativity and its shape, the molecule possesses a permanent dipole. The bond in hydrogen chloride has about 20 per cent ionic character.

When hydrogen combines with elements of low electronegativity, e.g. the alkali metals and the alkaline earths, it gains an electron (reduction) to form an ionic bond; for example, with sodium,

$$[Na\ 1s^2\ 2s^2\ 2p^6]^+ \quad [H\ 1s^2]^-$$
$$\text{sodium ion} \qquad \text{hydride ion}$$

The sodium ion has a 'neon' structure and the hydride ion a 'helium' structure. It is theoretically possible for the hydrogen ion to undergo the change

$$H - e \longrightarrow H^+$$

but the energy required for this process ($1310\ kJ\ mol^{-1}$) is so high that it seldom occurs. The ion H^+ is never found in chemical systems except perhaps as a transient species in certain chain reactions. It occurs in appreciable quantities only in hydrogen discharge tubes. The positive hydrogen ion is a powerful electron pair acceptor (Lewis acid) so that in aqueous solutions of, say, the mineral acids the 'hydroxonium' ion is present in relatively high concentration. For example, when hydrogen chloride gas is dissolved in water, hydroxonium ions are formed:

The hydroxonium ion can then react with strong bases such as the hydroxyl ion, by donating a proton:

$$(H_3O)^+ + OH^- \longrightarrow 2H_2O$$

The presence of the hydroxonium ions accounts for the 'acidic' properties of the solution. The above change is sometimes represented, *as a matter of convenience*, as

$$H^+ + OH^- \longrightarrow H_2O$$

but it must be remembered that 'H^+' is never present as such.

When hydrogen is covalently bonded to one of the three most electronegative elements, fluorine, oxygen or nitrogen, it can form a unique type of bond known as the hydrogen bond (as discussed for water in Chapter 3). For example, hydrogen fluoride exists as a liquid at room temperature because of hydrogen bonding between the HF molecules. The highly electronegative fluorine atom pulls the bonding pair towards itself, leaving a partially 'exposed' proton at the other end of the molecule. The 'exposed' proton then interacts with the electron field of a fluorine atom in another molecule of hydrogen fluoride to form a hydrogen bond:

hydrogen bond

The degree of polymerization will depend upon the temperature, since at higher temperatures the thermal vibration and motion of the molecules increases, and this will oppose hydrogen bond formation. The HF polymer is a 'zig-zag' molecule, since the hydrogen bond will form in the direction of maximum electron density, i.e. along the axis of a lone pair, and the lone and bonding pairs have the expected tetrahedral orientation.

Those d-block elements which combine with hydrogen do so to form non-stoichiometric hydrides, which are less dense than the parent metals. They are interstitial compounds, i.e. the hydrogen atoms fit into the interstices between the metal atoms in the crystal lattice. The very small size of the hydrogen atom is one factor which makes this possible.

The position of hydrogen in the Periodic Table

Hydrogen has been placed in both Groups IM and VIIM in the Periodic Table. The justification for this appears to be that (1) it formally resembles the alkali metals in that it has one s electron in its valence shell, and (2) it resembles the lighter halogens in that it is a gas at room temperature, shows little tendency to form a positive ion, and shows some tendency to form a negative hydride ion analogous to a halide ion. A more detailed comparison of hydrogen with the alkali metals and halogens is given in Table 7.2.

Property	Alkali metals	Hydrogen	Halogens
Electronegativity (Pauling)	Fr, 0·7 to Li, 1·0	2·1 →	I, 2·5 to F, 4·0
First ionization energy (kJ mol^{-1})	Cs, 377 to Li, 579	1310 →	I, 1008 to F, 1682
Covalent radius (nm)	Cs, 0·225 to Li, 0·134	0·037 ↓	I, 0·133 to F, 0·072
Ionic radius (+1) (nm)	Fr, 0·176 to Li, 0·06	very small ↓	—
Ionic radius (−1) (nm)	—	0·208 →	I, 0·216 to F, 0·136
Electron affinity (kJ mol^{-1})	Na, 72	← 67·4	I, 312·5 to F, 348·9
Electronic structure of valence shell	ns^1	← 1s^1	ns^2np^5

(Arrow points to group that has greater resemblance to hydrogen. Vertical arrow means that hydrogen shows little resemblance to either group.)

Table 7.2

Hydrogen has the following unique properties:

(*a*) It has a very small covalent radius, so, unlike either the alkali metals or the halogens it readily forms interstitial compounds with the transition metals.

(*b*) It has a very small +1 ionic radius, a large first ionization energy and a small electron affinity. It has, therefore, hardly any tendency to form an isolated H^+ ion and only a small tendency to form an H^- ion and this only with elements of very low electronegativity. In most of its compounds it forms covalent bonds.

(c) In the hydrogen atom there are no underlying shells between the nucleus and the valence shell. This enables the hydrogen to form hydrogen bonds when it is combined with elements of high electronegativity.

Because of these unique properties it is better not to place hydrogen in either Group I or Group VII, as shown in the Periodic Table given on p. 22.

Isotopes of hydrogen

There are three isotopes of hydrogen: ^1H, ^2H (deuterium) and ^3H (tritium). ^1H has one proton in its nucleus, ^2H a proton and a neutron and ^3H a proton and two neutrons. The relatively big percentage difference in mass between atoms of these isotopes leads to larger differences in physical properties of the isotopes and their compounds than is usually observed between isotopes of other elements; e.g. the boiling point of $^1H_2$ is $-252\cdot6°C$ while that of deuterium is $-249\cdot4°C$. They differ chemically in that analogous chemical reactions involving different isotopes have different equilibrium and rate constants.

Ordinary water contains a small percentage of deuterium oxide. When water is electrolysed ^1H is liberated more rapidly than ^2H so that the residual electrolyte becomes progressively richer in deuterium oxide (heavy water) as electrolysis proceeds. This gives a method of preparing heavy water. The physical properties of water and heavy water are shown below:

	H_2O	D_2O
Density at 20°C (kg m^{-3})	917	1017
Freezing point (°C)	0	3·82
Boiling point (°C)	100	101·42
Temperature of maximum density (°C)	4	11·6

2. The s-block elements

THE ALKALI METALS

Structural characteristics

(i) They all have one s electron in their valence shells.
(ii) The first ionization energies are all relatively low and decrease down the group.
(iii) The electronegativities are all small and decrease down the group.
(iv) The +1 ionic radii are relatively large and increase down the group.

Symbol	Electronic structure	First ionization energy (kJ mol^{-1})	Electro-negativity (Pauling)	RADII (nm)		
				Ionic (+1)	Covalent	Metallic (coordina-tion number 12)
Li	core [He] 2s^1	579	1·0	0·060	0·134	0·155
Na	[Ne] 3s^1	498	0·9	0·095	0·154	0·190
K	[Ar] 4s^1	418	0·8	0·133	0·196	0·235
Rb	[Kr] 5s^1	402	0·8	0·148	0·211	0·248
Cs	[Xe] 6s^1	377	0·7	0·169	0·225	0·267
Fr	[Rn] 7s^1	—	0·7	0·176	—	—

Table 7.3

(v) The covalent and metallic radii are large and increase from lithium to caesium.

(vi) All the alkali metal atoms, except lithium, have a penultimate shell of eight electrons.

General chemistry

The relatively small value of the first ionization energy and the large number of vacant energy levels in the valence shell result in the Group I elements being metals. They are malleable, ductile, lustrous, good conductors of electricity and so on. However, because of their large atomic volumes they are soft and light.

The low values of the first ionization energies and electronegativities result in a tendency to undergo the change

$$X - e \longrightarrow X^+$$

relatively easily. The resulting X^+ ion is very stable and, indeed, much of the chemistry of the Group I elements is the chemistry of these monovalent unipositive ions. The alkali metals readily form ionic bonds with the more electronegative elements, e.g. Na^+Cl^-, $Na^+O^{2-}Na^+$, and also with elements of moderate electronegativity, e.g. sodium hydride, Na^+H^-. In reactions of alkali metal compounds the monovalent metal cation remains unchanged throughout the reaction (except in some electrode processes), e.g.

$$Na^+ + OH^- + H^+ + Cl^- \longrightarrow Na^+ + Cl^- + H_2O$$

$$2Na^+ + CO_3^{2-} + Ba^{2+} + 2Cl^- \longrightarrow \downarrow BaCO_3 + 2Na^+ + Cl^-$$

$$Na^+ + Cl^- + H_2SO_4 \longrightarrow Na^+ + HSO_4^- + \uparrow HCl$$

The alkali metal cation, once formed, shows little tendency to regain its valence electron and therefore has very little distorting effect on an anion associated with it. The anion, therefore, has little tendency to decompose to give a more compact, stabler ion and so is relatively stable to heat at the temperature of a hot bunsen flame. Group I carbonates and hydroxides do not decompose on heating (except for lithium); the nitrates (apart from lithium) decompose only to the nitrite; e.g.

$$NaNO_3 \longrightarrow NaNO_2 + \tfrac{1}{2}O_2$$

Again, all the alkali metals except lithium form peroxides and higher oxides; e.g.

$$2Na^+ \; \left[:\ddot{O}:\ddot{O}: \right]^{2-} \; \text{or} \; 2Na^+ \; (O—O)^{2-} \quad \text{sodium peroxide}$$

$$Na^+ \; \left[\cdot\ddot{O}:\ddot{O}\cdot \right]^- \; \text{or} \quad Na^+ \; (O{=}O)^- \quad \text{sodium superoxide}$$

As well as a normal covalent bond between the oxygen atoms, the superoxide ion probably contains a 'one electron' bond as indicated by the broken line above.

An important factor governing the ability of a given ion to form complexes is the ionic radius. The smaller the ionic radius the greater the polarizing power of the ion, i.e. the greater its tendency to attract around itself electron-donating groups or molecules. Since the alkali metal cations have relatively large ionic radii, they have little tendency to attract such groups and so form complexes. This is reinforced by the low charge on the alkali metal ions, since the polarizing power increases with ionic charge. Thus, none of the alkali metal ions forms hydrated ions of definite formula in aqueous solution, though all of them are solvated to some extent.

The smallest alkali metal cation, Li^+, has the greatest polarizing power, and this shows the greatest degree of hydration in crystals of its salts. The degree of hydration then decreases going down the group, none of the salts of caesium and rubidium being hydrated. It might be expected that the mobilities of the alkali metal ions in aqueous solution would decrease from lithium to caesium, as ionic size increases. In fact, the mobilities increase from lithium to caesium. This is because the increase in ionic size is more than balanced by the decrease in degree of hydration; i.e. caesium has the greatest mobility because it has the smallest *hydrated* cation.

The fact that the alkali metals relatively easily undergo the process

$$X \, (g) - e \, (g) \longrightarrow X^+ \, (g)$$

means that the reverse process is achieved with difficulty. In order to extract the alkali metals from their compounds, therefore, the most vigorous form of reduction possible must be used, i.e. electrolysis of a fused compound. Sodium is prepared on a large scale by the electrolysis of the fused hydroxide or chloride.

Although most of the chemistry of the alkali metals is the chemistry of their cations, they do form covalent bonds, e.g. in lithium and sodium alkyls such as $Li-CH_3$, $Na-C_2H_5$.

The atypical properties of Period 2 elements: lithium and magnesium—a diagonal relationship

The elements of Period 2 (Li to F), which are the first elements in each of the main groups, exhibit properties which are not typical of the groups to which they belong. They differ structurally from the other elements in the groups in possessing only two s electrons in their penultimate shells. Whereas the valence electrons of the other elements in the group are shielded by at least eight penultimate electrons, the first element has only

Other alkali metals	Lithium	Magnesium
(1) b.p. Na, 892°C Cs, 690°C	b.p. 1330°C	b.p. 1107°C
(2) On heating, the nitrates decompose to the nitrite	On heating, the nitrate decomposes to the oxide	On heating, the nitrate decomposes to the oxide
(3) Soft metals	Much harder than other alkali metals	Much harder than alkali metals
(4) Form peroxides and higher oxides	Does not form a peroxide or higher oxide	Does not form a peroxide or higher oxide
(5) Their salts show a small tendency to be hydrated	Most salts are hydrated	Most salts are hydrated
(6) Form soluble carbonates	Carbonate insoluble	Carbonate insoluble
(7) Hydroxides are very soluble	Hydroxide sparingly soluble	Hydroxide sparingly soluble
(8) Hydroxide and carbonates do not decompose on heating	Hydroxide and carbonates decompose to the oxide on heating	Hydroxide and carbonate decompose to the oxide on heating
(9) Do not form nitrides	Forms a nitride, Li_3N	Forms a nitride, Mg_3N_2

Table 7.4

two shielding electrons. This has at least some of the following consequences:

(i) The first element in the group has a comparatively small atomic radius.
(ii) It has a small ionic radius compared to the other ions in the group and the same ionic charge, so that the ion has a relatively large polarizing power.
(iii) There is a bigger difference in (a) electronegativity, (b) first ionization energy, between the first and second element than between any other two consecutive elements in the group.

For the first four groups the *first* element in a group tends to resemble the *second* element in the next group. This is the so-called *diagonal relationship*. The polarizing power of an ion increases as the charge on the ion increases, and decreases as the ionic radius increases. Across a period the ionic charge increases while down a group the ionic radius increases. Diagonally, therefore, these two effects tend to cancel each other so that the polarizing power is approximately the same for ions of elements such as lithium and magnesium. This gives rise to certain similarities in properties—a diagonal relationship. After Group IV the significance of the diagonal relationship decreases as ionic character becomes less important.

The atypical properties of lithium and its diagonal relationship with magnesium are illustrated in Table 7.4.

THE ALKALINE EARTH METALS

Sym-bol	Electronic structure	IONIZATION ENERGIES (kJ mol^{-1})			Electro-negativity (Pauling)	RADII (nm)		
		1st	*2nd*	*1st + 2nd*		*Ionic (+2)*	*Cova-lent*	*Metallic (coordination number 12)*
Be	He 2s²	900	1753	2653	1·5	0·031	0·090	0·112
Mg	Ne 3s²	736	1448	2184	1·2	0·065	0·130	0·160
Ca	Ar 4s²	590	1146	1736	1·0	0·099	0·174	0·197
Sr	Kr 5s²	548	1059	1607	1·0	0·113	0·192	0·215
Ba	Xe 6s²	502	958	1460	0·9	0·135	0·198	0·222
Ra	Rn 7s²	—	—	—	0·9	0·140	—	—

Table 7.5

Structural characteristics

(i) They all have two s electrons in their valence shells.
(ii) The Group II elements have eight electrons in their penultimate shells, except for beryllium which has only two.

(iii) The first ionization energies are all relatively low and decrease rapidly down the group.

(iv) To form a dipositive ion both valence electrons must be lost, so that the sum of the first and second ionization energies must be considered. This sum is always more than double the ionization energy of the corresponding alkali metal, but decreases rapidly down the group.

(v) Apart from beryllium they all have relatively large ionic, covalent and metallic radii and these increase down the group.

(vi) The alkaline earth atoms have a relatively small number of valence electrons compared with the number of available sub-levels in the valence shell.

(vii) Their electronegativities are low and decrease down the group.

General chemistry (excluding beryllium, see p. 111)

In Group II the conditions for metallic bonding are fulfilled; that is, low first ionization energies (below 1050 kJ mol^{-1}) and a relatively large number of empty sub-levels in the valence shell. The alkaline earth elements, therefore, show many of the properties typical of metals, e.g. metallic lustre, good electrical and thermal conductivity.

Because of their low ionization energies and electronegativities, they also exhibit many of the chemical properties associated with metals. They readily lose their valence electrons to more electronegative elements and form dipositive ions. The chemistry of their simple compounds is largely the chemistry of their ions. Like the alkali metal ions, in many reactions these ions remain unchanged; e.g.

$$Ca^{2+} + 2OH^- + 2H^+ + 2Cl^- \longrightarrow Ca^{2+} + 2Cl^- + 2H_2O$$

$$Ba^{2+} + 2Cl^- + 2H^+ + SO_4^{2-} \longrightarrow Ba^{2+}SO_4^{2-} + 2H^+ + 2Cl^-$$

The sum of the first and second ionization energies is, however, much larger than the first ionization energy of the corresponding alkali metal, so that the change

$$X(g) - 2e(g) \longrightarrow X^{2+}(g)$$

takes place less readily than the change

$$X(g) - e(g) \longrightarrow X^+(g)$$

for the alkali metals. This means that the alkaline earth ions are a little easier to reduce than the alkali metal ions; i.e. they show a greater tendency to regain their valence electrons. In their compounds, therefore, the Group II ions have a greater distorting effect upon the ions with which they are associated than do the alkali metal ions. This means that compounds are more readily decomposed by heat, so that unlike the alkali metals (apart from Li) the nitrates, carbonates and hydroxides all decompose to the oxide on heating. Compared with the compounds of the metals lower in the electrochemical series such as copper, however, the temperatures of decomposition are quite high; e.g. $CaCO_3$ decomposes to the oxide at about 1000°C, while $CuCO_3$ is decomposed readily at 300–400°C.

The relatively high stability of the alkaline earth ions means that extraction of the metals from their compounds requires the most vigorous form of reduction, viz. electrolysis of a fused salt (usually the chloride).

The Group II ions differ from the Group I ions in the following important respects:

(1) They have twice the positive charge.
(2) They have relatively smaller ionic radii than alkali metals in the same period.

The alkaline earth ions, therefore, have a greater polarizing power, i.e. show a greater tendency to attract electrons from electron-donating groups. The ability to form complexes, although still not well developed, is thus slightly enhanced. (When a simple ion attracts around itself electron-donating groups or molecules the ion so formed is called a *complex*.) Magnesium forms a few complex cations, such as that shown in Fig. 7.1.

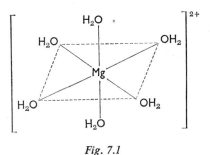

Fig. 7.1

Calcium forms a hexahydrate similar to the magnesium complex. As the ionic size of the ions increases going down Group II, the tendency to form complexes decreases.

A comparison of the properties of Group I and Group II elements is given in Table 7.6.

The anomalous properties of beryllium

Like the other elements in Period 2, beryllium shows anomalous properties. It has only two electrons in its penultimate shell and its valence

SIMILARITIES
(1) They have relatively low ionization energies and electronegativities.
(2) They form very stable cations.
(3) They are soft, light metals.
(4) Apart from Be and Mg they react readily with cold water.
(5) The metals are prepared from their chlorides by electrolysis.
(6) Apart from Be, Mg and Li they form peroxides.
(7) Both form hydrides which are ionic, apart from BeH_2 and MgH_2.

DIFFERENCES	
Alkaline earth metals	*Alkali metals*
(1) The carbonates and hydroxides break down to the oxides on heating.	Apart from those of Li, the carbonates and hydroxides are thermally stable.
(2) The nitrates decompose to the oxides on heating.	Apart from lithium nitrate, the nitrates decompose to the nitrite only on heating.
(3) They show some tendency to form complexes.	Apart from lithium, they show little tendency to form complexes.
(4) (a) Their hydroxides are not as soluble as the alkali metal hydroxides, but their solubility increases going down the group.	All their simple compounds are soluble in water.
(b) All their carbonates are insoluble.	
(c) Magnesium sulphate is soluble, but the other sulphates become increasingly insoluble going down the group.	
(5) Form nitrides X_3N_2	Only lithium forms a nitride.
(6) Form carbides $X^{2+}(C\equiv C)^{2-}$	Do not form carbides.

Table 7.6   Comparison of alkali metals and alkaline earth metals

shell can hold a maximum of eight electrons. Consequently, compared to the other alkaline earth metals, it has a relatively high electronegativity (1·5) and first ionization energy (900 kJ mol^{-1}). It has a very small ionic radius (0·031 nm) and this combined with the relatively high ionic charge (2+) gives it a high polarizing power. When the Be^{2+} ion forms a complex ion it cannot accept more than four electron pairs, because it is saturated with eight electrons in its valence shell, since the second main energy level can contain a maximum of only eight electrons.

The above structural characteristics lead to the following anomalous properties of beryllium.

(1) It has a much greater tendency to form covalent compounds than the other alkaline earths. For example, although the other Group II elements form ionic chlorides of the type $X^{2+}2Cl^-$, beryllium forms a covalent chloride $BeCl_2$. This is a linear molecule in the monomeric form, at high temperatures ($1750°C$):

$$Cl—Be—Cl$$
$$\overset{\longleftarrow 180° \longrightarrow}{}$$

but because of the high electron accepting power of the beryllium atom it tends to polymerize at lower temperatures, as shown in Fig. 7.2. Each beryllium atom is tetrahedrally coordinated by four

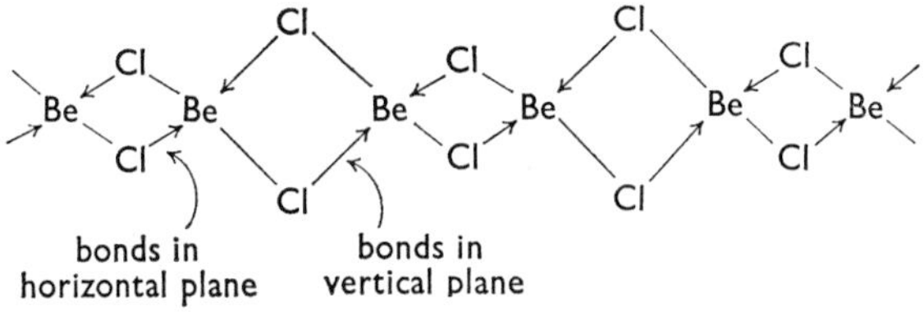

bonds in   bonds in
horizontal plane   vertical plane

Fig. 7.2

chlorine atoms. The tetrahedra are distorted, however, the bond angle being about $98°$ compared to the value for a regular tetrahedron, $109° \, 28'$.

(2) The Be^{2+} ion has high polarizing power, so it forms complexes readily; for example, $[BeF_4]^{2-}$ and $[Be(H_2O)_4]^{2+}$, in which the electron-donating groups (ligands) are tetrahedrally disposed about the central beryllium atom.

(3) The metal is not attacked by water because of the formation of a tenacious oxide film on its surface. It is rendered passive by nitric acid and is not very readily attacked by the other mineral acids.

(4) Beryllium, unlike the other alkaline earths, is amphoteric. It dissolves in acids to give a salt and hydrogen, and in alkalis to give a 'beryllate' and hydrogen; e.g.

$$Be + 2H^+ + 4H_2O \longrightarrow [Be(H_2O)_4]^{2+} + H_2$$

$$Be + 2OH^- + 2H_2O \longrightarrow [Be(OH)_4]^{2-} + H_2$$
$$\underset{\substack{\text{'beryllate'} \\ \text{ion}}}{}$$

(5) Beryllium carbonate is very unstable and can be preserved only in an atmosphere of carbon dioxide. It is hydrolysed immediately in aqueous solution to the basic carbonate.

(6) Beryllium forms a carbide of formula Be_2C which is hydrolysed by water to give methane:

$$Be_2C + 2H_2O \longrightarrow CH_4 + 2BeO$$

Beryllium and aluminium: a diagonal relationship

The ionic radius of Be^{2+} (0·031 nm) is considerably smaller than that of Al^{3+} (0·050 nm), but the aluminium ion has a triple positive charge so that the polarizing powers of the two ions are very similar. The electronegativities of the two elements are the same. They show, therefore, a number of similarities in properties; i.e. there is a diagonal relationship between them. The properties of aluminium and beryllium are compared in Table 7.7.

Beryllium	*Aluminium*
(1) b.p. 2770°C.	(1) b.p. 2450°C.
(2) BeO has a very high melting point, is non-volatile and hard.	(2) Al_2O_3 has a very high melting point, is non-volatile and hard.
(3) Be metal is not attacked by cold water.	(3) Al metal is not attacked by cold water.
(4) Be metal is rendered passive by nitric acid.	(4) Al metal is rendered passive by nitric acid.
(5) $BeCl_2$ is covalent.	(5) $AlCl_3$ is covalent.
(6) Carbonate is unstable.	(6) Carbonate is unstable.
(7) Be metal is amphoteric. $Be + 2H^+ + 4H_2O$ $\longrightarrow [Be(H_2O)_4]^{2+} + H_2$ $Be + 2OH^- + 2H_2O$ $\longrightarrow [Be(OH)_4]^{2-} + H_2$	(7) Al metal is amphoteric. $Al + 3H^+ + 6H_2O$ $\longrightarrow [Al(H_2O)_6]^{3+} + \frac{3}{2}H_2$ $Al + 3OH^- + 3H_2O$ $\longrightarrow [Al(OH)_6]^{3-} + \frac{3}{2}H_2$
(8) Be^{2+} readily forms complexes, e.g. $Be(H_2O)_4^{2+}$	(8) Al^{3+} readily forms complexes, e.g. $Al(H_2O)_6^{3+}$
(9) Beryllium forms a carbide Be_2C which reacts with water to give methane: $Be_2C + 2H_2O \longrightarrow CH_4 + 2BeO$	(9) Aluminium forms a carbide Al_4C_3 which reacts with water to give methane: $Al_4C_3 + 6H_2O \longrightarrow 3CH_4 + 2Al_2O_3$

Table 7.7

Magnesium, zinc and beryllium

The magnesium atom has two electrons in its valence shell and eight electrons in its penultimate shell. The zinc atom has two valence electrons and eighteen electrons in its penultimate shell. Zinc and magnesium, therefore, are structurally similar in that

(*a*) they both have two valence electrons,
(*b*) they both have completed penultimate shells.

The relative positions, in the Periodic Table, of zinc and magnesium are such that the effect on atomic size of the greater nuclear charge of zinc is approximately balanced by its greater number of electrons, so that the ionic, covalent and metallic radii of the two metals have similar values. These values are shown in Table 7.8. They result in a similarity in chemical properties between magnesium and zinc, as shown in Table 7.9.

	Magnesium	*Zinc*
Structure	Mg $1s^2\ 2s^2\ 2p^6\ 3s^2$ completed shell	Zn $1s^2\ 2s^2\ 2p^6\ 3s^2\ 3p^6\ 3d^{10}\ 4s^2$ completed shell
Ionic radius (nm)	0·065 (+2)	0·074 (+2)
Covalent radius (nm)	0·130	0·131
Metallic radius (nm)	0·160	0·138

Table 7.8

Calcium *Strontium* *Barium*	*Magnesium*	*Zinc*
(1) Liberate hydrogen from cold water.	Does not liberate hydrogen from cold water.	Does not liberate hydrogen from cold water.
(2) Form peroxides.	Does not form a peroxide.	Does not form a peroxide.
(3) Form insoluble sulphates.	Forms a soluble sulphate, $MgSO_4.7H_2O$	Forms a soluble sulphate, $ZnSO_4.7H_2O$

Table 7.9

Although there are some similarities between zinc and magnesium (indeed, in some ways magnesium resembles zinc more than the other members of its group), there are also considerable differences. The sums of the first and second ionization energies for each element are quite different and so are the electronegativities. In these respects zinc shows a closer resemblance to beryllium. These properties are shown for beryllium, zinc and magnesium in Table 7.10.

	Beryllium	Zinc	Magnesium
Electronegativity	1·5	1·6	1·2
Sum of 1st and 2nd ionization energies (kJ mol^{-1})	2657	2640	2184

Table 7.10

Zinc resembles beryllium chemically in certain respects, e.g. they are both amphoteric, forming 'zincates' and 'beryllates'. Again, they both form complex ions in which the central metal ion is surrounded tetrahedrally by four ligands, e.g. $Be(OH)_4^{2-}$ and $Zn(OH)_4^{2-}$.

References

Books

Concise Inorganic Chemistry, J. D. Lee (Van Nostrand)
Inorganic Chemistry: An Advanced Textbook, T. Moeller (Wiley)
Advanced Inorganic Chemistry, F. A. Cotton and G. Wilkinson (Wiley)
Comparative Inorganic Chemistry, B. J. Moody (Arnold)

8. The p-block elements

Group IIIM

Sym-bol	Electronic structure	Ionization energy (kJ mol^{-1})		Electro-negativity (Pauling)	RADII (nm)		
		1st	1st + 2nd + 3rd		Ionic	Cova-lent	Metallic
B	(He) 2s^2 2p^1	799	6761	2·0	0·020 (+3)	0·082	0·098
Al	(Ne) 3s^2 3p^1	577	5112	1·5	0·050 (+3) 0·148 (+1)	0·118	0·143
Ga	(Ar) 3d^{10} 4s^2 4p^1	577	5497	1·6	0·062 (+3) 0·132 (+1)	0·126	0·141
In	(Kr) 4d^{10} 5s^2 5p^1	557	5067	1·7	0·081 (+3) 0·140 (+1)	0·144	0·166
Tl	(Xe) 4f^{14} 5d^{10} 6s^2 6p^1	590	5404	1·8	0·095 (+3)	0·148	0·171

Structural characteristics

(i) All the elements of this group have completed inner shells and a valence shell with the structure $ns^2 np^1$.

(ii) Gallium, indium and thallium have valence shells that are directly underlain by completed d sub-levels which have a relatively poor screening effect. The completed 5d level of thallium is underlain by a completed 4f sub-level which again has a relatively poor screening effect. From gallium to thallium, therefore, the increase in nuclear charge (atomic number) will be the important factor affecting the attraction of peripheral electrons.

(iii) The electronegativity decreases from boron to aluminium and then slowly increases to thallium.

(iv) The Group IIIM elements have relatively low first ionization energies, which decrease from boron to indium and increase at thallium (big increase in nuclear charge).

    The sums of the first, second and third ionization energies vary rather irregularly going down the group, as shown in Fig. 8.1. There is a relatively large decrease from boron to aluminium because of a relatively large increase in the number of inner electrons and their screening power. There is then an increase from aluminium to gallium because the relatively large increase of atomic number by eighteen units is the important factor.

The ionization energy then decreases from gallium to indium as the number and screening power of the inner electrons increases. Finally, there is an increase from indium to thallium as the atomic number increases by 32 units, the additional fourteen 4f electrons having relatively poor screening power.

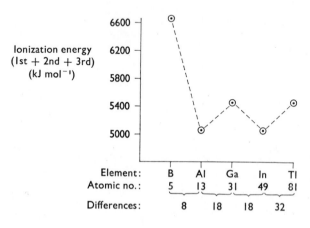

Fig. 8.1

(v) The elements of Group IIIM have relatively small $+3$ ionic radii and relatively large $+1$ ionic radii, and these increase going down the group. The covalent radii increase going down the group.

(vi) The s electrons in the valence shells of the Group IIIM elements show some tendency to remain paired and not take part in bonding. This tendency increases going down the group as the atomic number increases; e.g. Tl(I) (see below). This is known as the *inert pair* effect.

The general chemistry of Group IIIM

The elements aluminium to thallium are silvery white metals. They are malleable, ductile and good conductors of heat and electricity. They all form metal (III) oxides of the type X_2O_3. Boron (III) oxide is acidic, aluminium (III) and gallium (III) oxides are amphoteric, and those of indium and gallium are basic.

The inert pair effect increases going down the group, so that boron always has an oxidation number of $+3$, aluminium nearly always has an oxidation number of $+3$ (though it is possible that aluminium (I) fluoride, AlF, exists), while gallium, indium and thallium show variable valency. They form two well-defined series of compounds in which the metal has oxidation numbers of $+1$ and $+3$ respectively. As might be expected, since the inert pair effect is greatest for the last element in the group, thallium (I) compounds are more stable than thallium (III) compounds. The ionic radius of the thallium (I) cation (0·140 nm) is similar to that of the rubidium cation (0·148 nm), so that thallium (I) hydroxide, TlOH, is a very strong base.

The relatively small ionic size and high charge on tripositive ions of this group means that the presence of M^{3+} ions in simple compounds is very unlikely. Thus, although the Al^{3+} ion is present in aluminium (III) fluoride, because of the high electronegativity of the fluorine, the other aluminium (III) halides are predominantly covalent. Aluminium (III) chloride, at high temperatures, has the structure:

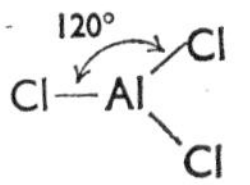

$$Cl-Al\overset{\overset{\displaystyle 120°}{\diagup}Cl}{\diagdown Cl}$$

The aluminium atom has a share in six electrons, but its valence shell can contain eight, so that at room temperature it dimerizes to give Al_2Cl_6 molecules, as shown in Fig. 8.2. The aluminium atoms then have

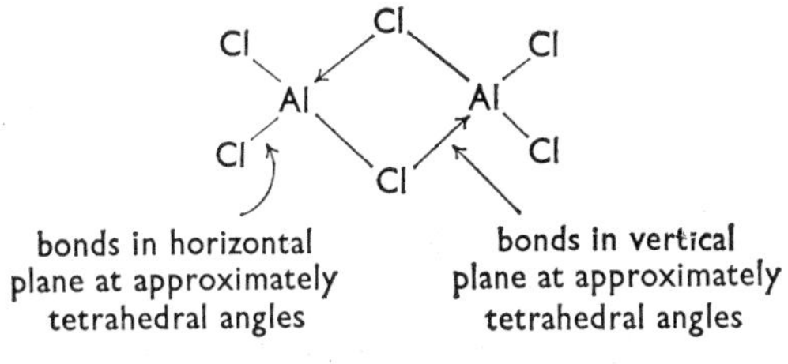

bonds in horizontal plane at approximately tetrahedral angles   bonds in vertical plane at approximately tetrahedral angles

Fig. 8.2

eight electrons in their valence shells. Anhydrous aluminium chloride is a solid made up of white silky crystals which fume in moist air, dissolve in organic solvents and are rapidly hydrolysed by water.

The B^{3+} ion is so small that even the fluoride has considerable covalent character:

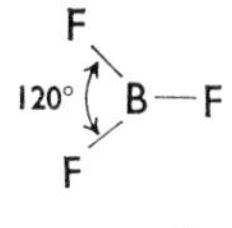

$$120°\overset{\displaystyle F}{\underset{\displaystyle F}{\Big(}}B-F$$

Since the boron atom can accommodate another electron pair in its valence shell, boron trifluoride is an electron-pair acceptor (Lewis acid) and readily forms addition compounds with electron-donating molecules (Lewis bases) such as ammonia, as shown in Fig. 8.3. Boron trifluoride does not form a dimer corresponding to Al_2Cl_6 because the monomer is stabilized by π-bonding between full p orbitals on the fluorine atom and the empty p orbital on the boron atom. This is shown in Fig. 8.3.

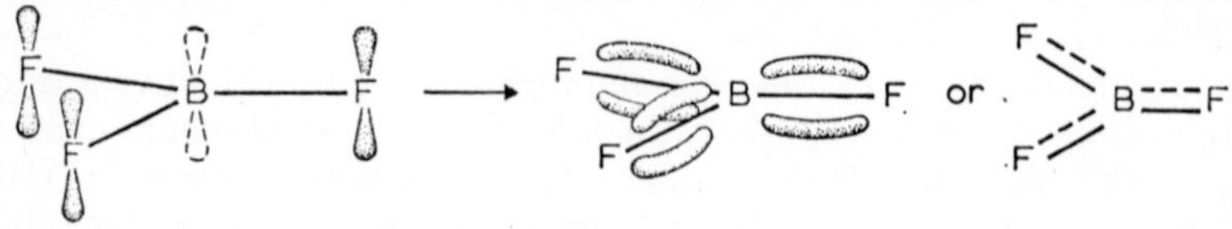

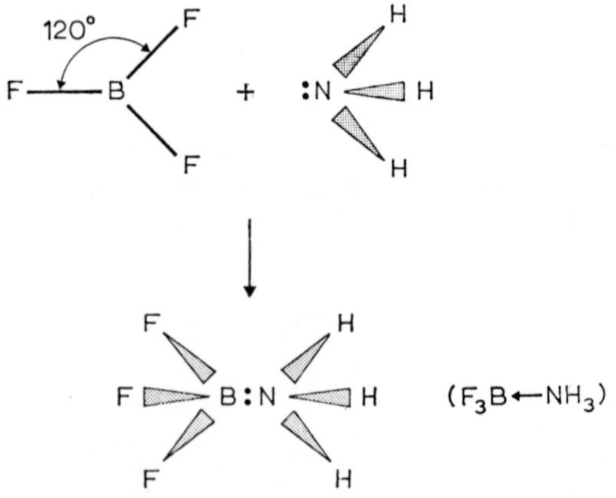

Fig. 8.3

Since the tripositive ions of the metals of Group IIIM have small ionic radii and relatively high charge they readily accept electrons from electron-donating groups and so form complex ions. Aluminium forms cationic and anionic complexes, as shown by the examples in Fig. 8.4.

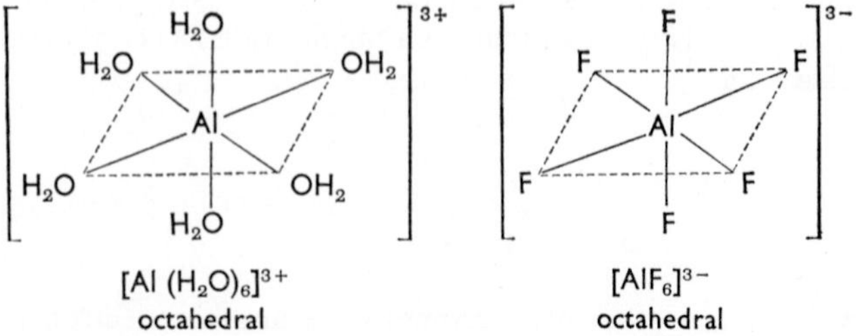

$[Al(H_2O)_6]^{3+}$
octahedral

$[AlF_6]^{3-}$
octahedral

Fig. 8.4

When anhydrous aluminium chloride is dissolved in water it is immediately hydrolysed, giving $[Al(H_2O)_6]^{3-}$ ions:

$$Al_2Cl_6 + 12H_2O \longrightarrow 2[Al(H_2O)_6]^{3-} + 6Cl^-$$

Aluminium metal, being amphoteric, dissolves in caustic soda solution to give complex 'aluminate' ions and hydrogen:

$$Al + 3OH^- + 3H_2O \longrightarrow [Al(OH)_6]^{3-} + \tfrac{3}{2}H_2$$

When caustic soda solution is added to an aqueous solution of an aluminium salt a gelatinous white precipitate of aluminium hydroxide is obtained which dissolves in excess to give a solution of sodium 'aluminate'. This reaction is best described in terms of the proton-donating capacity (acidic character) of the $[Al(H_2O)_6]^{3+}$ ion:

$$[Al(H_2O)_6]^{3+} \xrightarrow{\ OH^-\ } [Al(H_2O)_5(OH)]^{2+} \xrightarrow{\ OH^-\ } [Al(H_2O)_4(OH)_2]^+$$
$$\xrightarrow{\ OH^-\ } \underset{\substack{\text{white} \\ \text{gelatinous} \\ \text{precipitate}}}{[Al(H_2O)_3(OH)_3]^0}$$

$$[Al(H_2O)_3(OH)_3]^0$$
$$\xrightarrow{\ OH^-\ } [Al(H_2O)_2(OH)_4]^- \xrightarrow{\ OH^-\ } [Al(H_2O)(OH)_5]^{2-} \xrightarrow{\ OH^-\ } \underbrace{[Al(OH)_6]^{3-}}$$
$$\underset{\text{'aluminate' ions}}{}$$

The proton-donating tendency of the $[Al(H_2O)_6]^{3+}$ ion also accounts for the acidic properties of aqueous solutions of aluminium salts:

$$[Al(H_2O)_6]^{3+} + H_2O \rightleftharpoons [Al(H_2O)_5(OH)]^{2+} + H_3O^+$$

Although aluminium has relatively low electronegativity, it is not attacked directly by air or oxygen. This is because a tenacious oxide film forms on its surface, and protects the underlying metal. If the surface of the aluminium is amalgamated, to remove the oxide film, then attack by air occurs readily. Aluminium has, in fact, a high affinity for oxygen and can reduce the oxides of metals like chromium to the free metal, as in the 'thermit' process. The reaction once started is highly exothermic. Aluminium oxide can be reduced to the metal only by the most vigorous form of reduction, electrolysis of a solution of the oxide in fused cryolite, Na_3AlF_6.

The anomalous properties of boron

Boron is in Period 2 and as such shows behaviour which is not typical of the other elements of the group. Boron has only two electrons in its penultimate shell, whereas the other elements of the group have at least eight. Consequently boron has a much smaller covalent and +3 ionic radius than the other elements in the group. Again, its first ionization energy, the sum of the first three ionization energies, and its electronegativity, are relatively much higher. These anomalous structural characteristics result in a number of anomalous properties.

5+

Boron forms a well-defined series of hydrides,

$$B_2H_6, \quad B_4H_{10}, \quad B_5H_9, \quad B_5H_{11}, \quad B_6H_{10}, \quad B_9H_{15}, \quad B_{10}H_{14}, \quad B_{10}H_{16}$$

whereas aluminium forms a hydride $(AlH_3)_x$ of unknown structure.

The boron hydrides are known as boranes and are named according to the number of boron atoms present; e.g. B_2H_6 is known as diborane. The boranes are interesting because of their unusual structures. If boron combines with three hydrogen atoms this gives the structure:

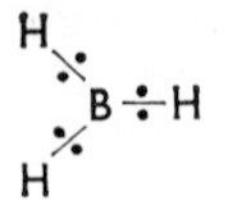

Since there are no lone pairs available on either the boron or the hydrogen atom it is difficult to see at first sight how a molecule such as B_2H_6 can be formed. Electron diffraction experiments indicate that the diborane molecule has the following structure and bond lengths:

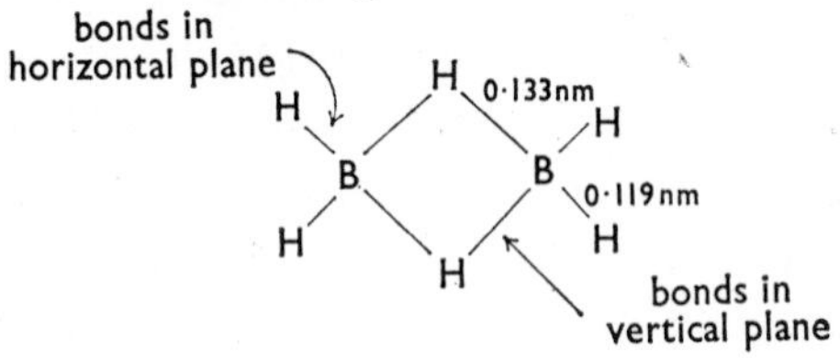

The boron atoms are held together by two hydrogen 'bridges', and the terminal hydrogens are bonded to the boron atoms by ordinary covalent bonds. This leaves one electron available on each boron atom and one on each of the remaining hydrogen atoms. The formation of the hydrogen 'bridges' is illustrated in Fig. 8.5. In each hydrogen bridge, two electrons are delocalized over three nuclei. For this reason the bond is sometimes called a three-centre bond. Diborane is said to have an *electron-deficient* molecule.

Diborane reacts with ammonia under certain conditions to give borazole, $B_3N_3H_6$. This has the structure illustrated in Fig. 8.6. It consists of a hexagonal ring of alternate boron and nitrogen atoms joined by σ-bonds. The hydrogen atoms are at 120° to the other bonds in the same plane. There is a delocalized π-bond which forms a 'bicycle tyre' of charge cloud above and below the plane of the ring. It is very similar in structure to benzene. In fact, the number of electrons in borazole is exactly the same as in benzene, i.e. benzene and borazole are *isoelectronic*. Consequently borazole shows many 'aromatic' properties.

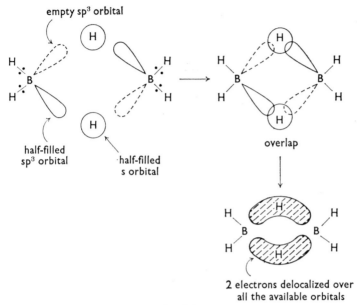

empty sp³ orbital

half-filled
sp³ orbital

half-filled
s orbital

overlap

2 electrons delocalized over
all the available orbitals

Fig. 8.5

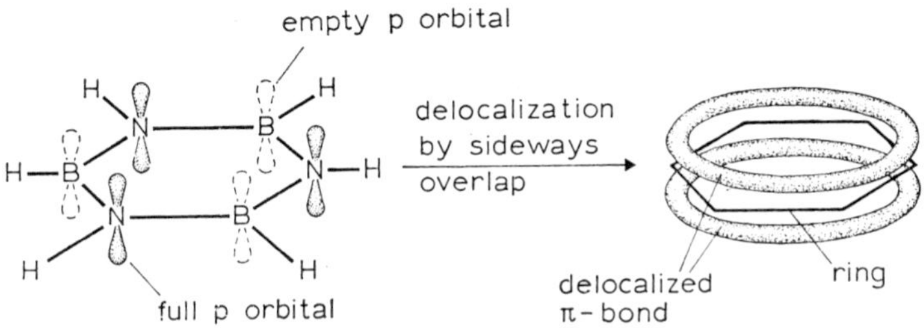

empty p orbital

delocalization
by sideways
overlap

full p orbital

delocalized
π−bond

ring

Fig. 8.6

As might be expected, boron nitride, BN, has a giant-molecule structure very similar to graphite. It is made up of macromolecular sheets of hexagonal rings, as shown in Fig. 8.7. A delocalized π-bond covers the whole of the sheet.

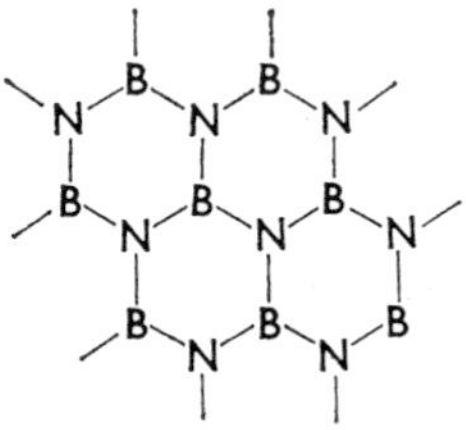

Fig. 8.7

As mentioned earlier, boron (III) oxide is an acidic oxide. It can give rise to orthoboric acid:

$$B_2O_3 + 3H_2O \longrightarrow 2H_3BO_3$$

which has the structure

This, on heating to 100°C, gives metaboric acid:

The orthoborate ion is trigonal in shape:

orthoborate ion, $(BO_3)^{3-}$

The metaborates on the other hand form chain or ring structures made up of triangular BO_3 units. These are illustrated in Fig. 8.8.

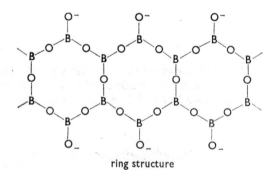

chain structure

ring structure

Fig. 8.8   Metaborate polymer

On dissolving in water the polymeric metaborate ions immediately break down to give $(BO_3)^{3-}$ units.

The anomalous properties of boron are compared to the properties of the rest of the group in Table 8.1.

Boron	Remaining elements in group
(1) Non-metallic physically.	(1) Definite metallic character.
(2) Forms a series of well-defined hydrides.	(2) Form hydrides with difficulty. These, when formed, are of uncertain structure.
(3) Forms compounds, such as boron nitride and borazole, which are analogous to the iso-electronic carbon compounds.	(3) Do not form compounds having carbon analogues.
(4) Halides are gaseous.	(4) Halides are solids.
(5) Boron (III) oxide is acidic.	(5) Metal (III) oxides are amphoteric or basic.
(6) Forms polymeric oxy-anions.	(6) Do not form polymeric oxy-anions.
(7) Melting point 2030°C.	(7) Melting points below 661°C.

Table 8.1

Although boron shows anomalous properties with respect to the other elements in Group IIIM it has a diagonal relationship with silicon, the second element in the next group. Some of the similarities between boron and silicon are summarized in Table 8.2.

Boron	Silicon
(1) A non-metal.	(1) A non-metal.
(2) High melting point (2030°C).	(2) High melting point (1410°C).
(3) Forms an acidic oxide, B_2O_3.	(3) Forms an acidic oxide, SiO_2.
(4) Forms polymeric oxy-anions.	(4) Forms polymeric oxy-anions.
(5) Gives rise to a series of covalent hydrides, e.g. B_2H_6.	(5) Gives rise to a series of covalent hydrides, e.g. Si_2H_6. (These are *not* electron-deficient structures.)

Table 8.2

Group IVM

Symbol	Electronic structure	IONIZATION ENERGY (kJ mol^{-1})		Electro-negativity (Pauling)	RADII (nm)		
		1st	1st+2nd+ 3rd+4th		Ionic (calcu-lated)	Cova-lent	Metallic
C	(He) $2s^2\,2p^2$	260	3415	2·5	0·260 (−4) 0·015 (+4)	0·077	—
Si	(Ne) $3s^2\,3p^2$	188	2375	1·8	0·271 (−4) 0·041 (+4)	0·111	—
Ge	(Ar) $3d^{10}\,4s^2\,4p^2$	187	2389	1·8	0·093 (+2) 0·053 (+4)	0·122	0·137
Sn	(Kr) $4d^{10}\,5s^2\,5p^2$	169	2115	1·8	0·112 (+2) 0·071 (+4)	0·141	0·162
Pb	(Xe) $4f^{14}\,5d^{10}\,6s^2\,6p^2$	171	2229	1·8	0·120 (+2) 0·084 (+4)	0·147	0·175

Structural characteristics

(i) They all have the structure $ns^2\,np^2$ for their valence shell.

(ii) The $+4$ ionic radii are small but increase down the group.

(iii) The covalent and metallic radii increase down the group.

(iv) The first ionization energies decrease down the group, except for lead which shows a small increase. The value for lead shows an increase because the additional fourteen 4f electrons which, although associated with an increase in nuclear charge of fourteen units, have a relatively poor screening effect.

(v) The electronegativity decreases from carbon to silicon and then remains constant at 1·8. This is probably because in atoms of the elements germanium to lead the screening effect of d and f electrons is relatively poor and is cancelled by the associated increase in nuclear charge.

(vi) Carbon is in Period 2 so that its valence shell can contain a maximum of eight electrons. When carbon has formed four covalent links it can accommodate no more electrons.

(vii) Each of the elements in Group IVM has two of its valence electrons in an s sub-level. As the atomic number increases (i.e. going down the group) these electrons tend to remain paired and not take part in bonding. The inert pair effect increases down the group.

(viii) The $+2$ ionic radius increases from germanium to lead.

General chemistry

Theoretically the donation of the four valence electrons of a Group IV element would lead to the formation of an M^{4+} ion with a relatively stable structure. However, such an ion would have a small ionic radius and a high charge, so its formation is very improbable. In general, therefore, all the simple quadrivalent compounds of elements in this group are covalent; e.g. the chlorides, CCl_4, $SiCl_4$, $GeCl_4$, $SnCl_4$, $PbCl_4$.

Since the chemistry of the elements from Group IV to Group VII is mainly the chemistry of covalent compounds, diagonal relationships, e.g. between carbon and phosphorus, are of little significance.

The inert pair effect increases down the group so that carbon and silicon have a valency of 4 in nearly all their compounds, while tin and lead show variable valency, forming compounds in which they have oxidation numbers of $+2$ and $+4$ respectively. Germanium (II) and tin (II) compounds are strong reducing agents, tending to be oxidized to the metal (IV) state. The inert pair effect is at a maximum, however, with lead, so that lead (II) compounds are not reducing agents and in the majority of its common compounds lead is in the $+2$ oxidation state. The $+2$ ionic radii of germanium, tin and lead are larger than the corresponding $+4$ ionic radii, so that metal (II) compounds show correspondingly greater ionic character.

All the elements of Group IVM form hydrides of the type XH_4:

CH_4 methane
SiH_4 silane
GeH_4 germane
SnH_4 stannane
PbH_4 plumbane

Carbon forms a large number of hydrides, the hydrocarbons, and silicon forms a series of saturated hydrides as far as Si_6H_{14}, hexasilane. The hydrides of silicon, germanium, tin and lead are spontaneously inflammable in air and are readily hydrolysed by alkalis. The hydrides of carbon on the other hand are characterized by their resistance to attack by most chemical reagents. The stability of, say, methane compared to the other tetrahydrides is probably associated with two factors:

(a) Unlike the other elements in the group carbon is more electronegative than hydrogen, so that the charge distribution in the methane molecule is different from that in the other hydrides, thus:

charge distribution
in methane

charge distribution in
silane, germane, stannane
and plumbane

(b) The carbon atom is saturated when it has formed four single covalent links with the hydrogens, while the valence shells of the other Group IV metals, in these circumstances, can hold more than eight electrons; i.e. the hydrides of the remaining element can act as electron acceptors (Lewis acids). This provides a possible mechanism for attack by species such as OH^-.

The elements of this group all form monoxides: CO, SiO, GeO, SnO, PbO. Carbon monoxide is stable and neutral, while silicon monoxide is very unstable. The monoxides of the last three elements are stable and amphoteric, though lead monoxide shows predominantly basic characteristics. The structure of carbon monoxide may be represented by

$$C \rightleftharpoons O$$

where the carbon has a share of eight electrons. Carbon monoxide forms carbonyl compounds with certain transition elements, e.g. $Ni(CO)_4$, nickel carbonyl, in which the central nickel atom is surrounded tetra-hedrally by four CO groups:

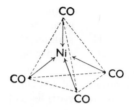

All the Group IVM elements form dioxides, as shown in Table 8.3.

Formula	Name	Properties	Salt (with NaOH)
CO_2	Carbon dioxide	Acidic gas	Na_2CO_3, sodium carbonate
SiO_2	Silicon dioxide or silica	Acidic solid of high m.p.	Na_2SiO_3, sodium silicate
GeO_2	Germanium dioxide or germanium (IV) oxide	Amphoteric solid	Na_2GeO_3, sodium germanate
SnO_2	Tin dioxide or tin (IV) oxide	Amphoteric solid	Na_2SnO_3, sodium stannate
PbO_2	Lead dioxide or lead (IV) oxide	Amphoteric solid	Na_2PbO_3, sodium plumbate

Table 8.3

Carbon dioxide and silica show an interesting contrast in physical properties at room temperature; carbon dioxide is a gas, while silica is a crystalline solid of very high melting point. This difference arises from a difference in structure. Carbon dioxide is made up of discrete, linear CO_2 molecules,

$$O=C=O$$

Intermolecular attraction arises from weak Van der Waals forces which are easily overcome, even at room temperature, by the thermal motion of the molecules, so that carbon dioxide is gaseous. On the other hand there are no discrete molecules of SiO_2 in silica. This has a giant-molecule structure, each crystal being made up of a network of cova-lently bonded atoms, in which each silicon atom is tetrahedrally sur-rounded by four oxygen atoms, as shown in Fig. 8.9. Because of these powerful inter-atomic forces, silica is an involatile crystalline solid. This difference in behaviour between carbon and silicon arises because silicon, like the other Period 3 elements, has empty 3d orbitals available. After silicon has formed four single covalent bonds with oxygen atoms in the quartz structure, the non-bonding electrons of the oxygen atoms

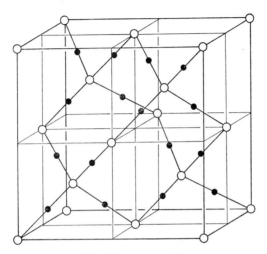

Fig. 8.9   The structure of quartz

can overlap with the empty d orbitals of the silicon atoms giving additional π-type bonding with a consequent increase in bond energy. The average bond energy for a C–O bond is 343 kJ mol^{-1} compared to that for Si–O bonding which is 368 kJ mol^{-1}. This additional π-bonding cannot occur when carbon has formed four covalent bonds, because the latter (being in Period 2) has no empty d orbitals available in its valence shell.

Since the inert pair effect is at a maximum with lead, lead dioxide is an oxidizing agent; e.g. it oxidizes hot concentrated hydrochloric acid to chlorine:

$$Pb^{4+} + 2Cl^- \longrightarrow Pb^{2+} \text{ (aq)} + Cl_2$$

All the elements of the group form tetrahalides of the type XY_4, where X is the Group IV element and Y a halogen, e.g. the tetrachlorides

$$CCl_4, \quad SiCl_4, \quad GeCl_4, \quad SnCl_4, \quad PbCl_4$$

These are all covalent liquids. CCl_4 is inert, for instance, with respect to water; the rest are readily hydrolysed.

The elements silicon to lead can accommodate more than eight electrons in their valence shells because in each case there are d sub-levels available, so that they are not saturated when four covalent bonds have

5*

been formed. This means that the halides of the elements silicon to lead are susceptible to attack by electron-donor molecules; e.g.

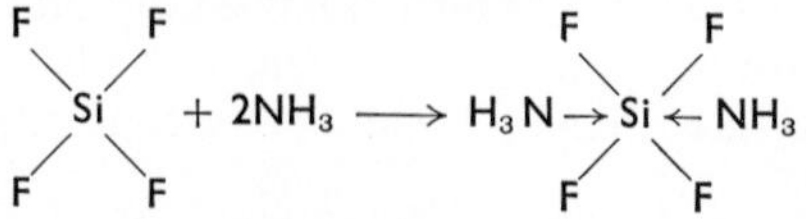

Consequently a possible mechanism for hydrolysis would be as follows:

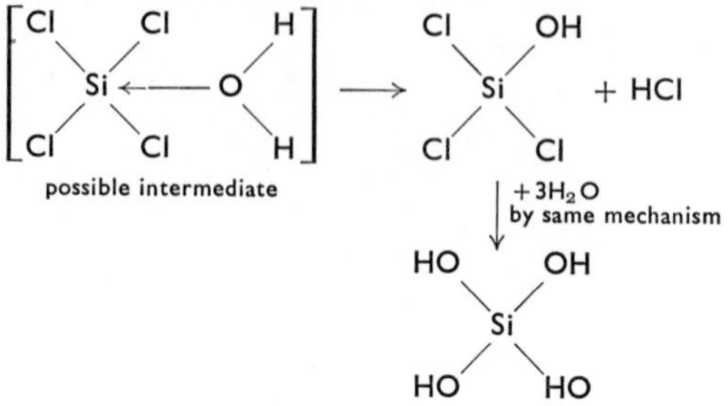

On the other hand there are no available sub-levels in the valence shell of the carbon atom in carbon tetrachloride, so that no addition compounds are possible. Carbon tetrachloride does not form addition compounds with ammonia and is not readily hydrolysed.

Diamond	*Graphite*
(1) Transparent (highly localized electrons and no absorption of incident visible radiation).	(1) Black with metallic lustre. (Mobile π-type electrons absorb incident radiation. Re-emission of part of the radiation as excited electrons return to ground state probably accounts for lustre.)
(2) Very involatile (strong bonds).	(2) Involatile; m.p. 3727°C, b.p. 4830°C (strong bonds).
(3) Bad conductor of electricity (no mobile electrons).	(3) Moderately good conductor of electricity (mobile π electrons).
(4) Very hard (every atom joined to four others by very strong covalent bonds).	(4) Soft and flaky (weak forces between sheets).
(5) Inert to attack by most reagents (strong bonds must be broken before reaction can occur).	(5) Is attacked by strong oxidizing agents to give 'graphitic oxide', a non-stoichiometric compound. (The π electrons are available for bonding. Spacing between layers is large enough for atoms to enter and form lamellar compounds. The distance between the layers increases when this occurs.)

Table 8.4

Carbon occurs in two allotropic forms, diamond and graphite; their structures have been discussed on p. 82. Silicon and germanium both adopt the diamond structure. Tin has two allotropes, grey tin which has the diamond structure, and white tin, a metallic form. Lead has a typically metallic cubic close-packed structure.

The differences in properties between diamond and graphite can be correlated with their different structures. A diamond crystal has a giant-molecule structure in which each carbon atom is joined to four tetrahedrally orientated neighbours by single covalent bonds. The bonding electrons are highly localized between the nuclei. Graphite, on the other hand, is made up of macromolecular sheets with weaker bonds holding the sheets together, each carbon being covalently bonded to three nearest neighbours in the same plane. The remaining electrons, not used for σ-bonding, are delocalized over the whole sheet. These structural differences lead to the differences in properties given in Table 8.4

The anomalous properties of carbon

Carbon, like the other Period 2 elements, tends to show anomalous properties with respect to the other elements in Group IV. It differs structurally from the other elements in the group in the following ways:

(a) Its valency shell is full when it contains eight electrons; i.e. the carbon atom is saturated when it has formed four single bonds.

(b) It is the most electronegative element in the group, and there is a much greater decrease in electronegativity from carbon to silicon than between any other two consecutive Group IV elements.

(c) Carbon has the smallest ionic and covalent radii in the group.

(d) It has the largest first ionization energy in the group, and there is a much bigger decrease from carbon to silicon than between any other consecutive Group IV elements.

(e) On the electronegativity scale, carbon lies approximately midway between fluorine (the most electronegative element) and caesium (the least electronegative element):

	Cs	C	F
electronegativity	0·7	2·5	4
difference		1.8	1·5

(f) At carbon, the inert pair effect is a minimum, so the tendency to exert a covalency of four is a maximum.

Associated with carbon's unusual structural characteristics are certain unusual properties. It can form stable single, double and triple bonds with itself, as in methane, ethylene and acetylene. Carbon atoms can bond with each other to give chains and rings, e.g. in the hydrocarbons. Carbon is unique in the extent of this chain-forming capacity and as a result its compounds outnumber those of the other elements. Again, no other element forms compounds as complex as many carbon compounds. Carbon compounds are the building blocks of living matter. Carbon can form stable compounds with elements of very high electronegativity and those of moderate electronegativity, e.g. CF_4 and CH_4. Carbon forms more stable double bonds with oxygen than does silicon. It combines with nitrogen to form the stable cyanide ion $(C \equiv N)^-$. This forms stable complexes with many of the transition elements, e.g. $[Fe(CN)_6]^{3-}$. The cyanide ion is similar in some ways to halide ions; for example, potassium cyanide reacts with copper (II) salts in a similar way to potassium iodide:

$$Cu^{2+} + I^- \longrightarrow Cu^+ + \tfrac{1}{2}I_2$$

$$Cu^{2+} + CN^- \longrightarrow Cu^+ + \tfrac{1}{2}\underset{\text{cyanogen}}{(CN)_2}$$

Cyanogen shows some similarity to the halogens and is sometimes called a pseudo-halogen. For instance, it reacts with dilute alkali in a similar manner to chlorine:

$$Cl_2 + 2OH^- \longrightarrow H_2O + Cl^- + ClO^-$$

$$(CN)_2 + 2OH^- \longrightarrow H_2O + CN^- + CNO^-$$

Carbon and silicon

The behaviour of silicon is in many ways quite different from that of carbon. Silicon–oxygen 'single' bonds are more stable than carbon–oxygen single bonds because silicon–oxygen bonds involve additional π-type bonding in which the non-bonding electrons of the oxygen atom overlap with empty d orbitals on the silicon atom as explained on p. 129. (Bond energies: Si–O = 368 kJ mol^{-1}; C–O = 343 kJ mol^{-1}.) Silicon–silicon bonds are much less stable than carbon–carbon single bonds. (Bond energies: Si–Si = 188 kJ mol^{-1}; C–C = 337 kJ mol^{-1}.) In contrast to carbon, therefore, silicon tends to form silicon–oxygen chains and rings rather than silicon–silicon structures. The tendency of silicon to form polymeric oxy-ions illustrates this point. These ions are present in various rocks and minerals. Examples are shown in Fig. 8.10. Electrical neutrality of the lattices containing the anions of Fig. 8.10 is preserved by the presence of the appropriate number of various metallic cations.

Three-dimensional network silicates may be formed by the sharing of all four oxygens with adjacent tetrahedra. This gives quartz, $(SiO_2)_n$. If some of the silicon atoms are replaced by Al^{3+} in tetrahedral positions and there are additional metal ions present to preserve electrical neutrality, this gives an important group of minerals known as *feldspars*; e.g. $KAlSi_3O_8$, orthoclase, and $NaAlSi_3O_8$, albite.

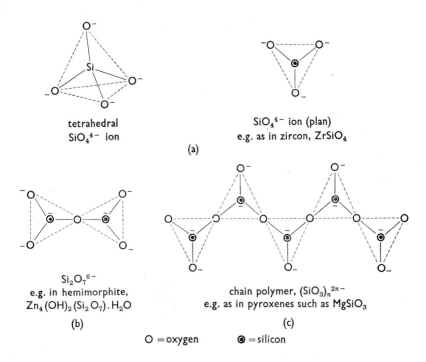

tetrahedral
SiO_4^{4-} ion

SiO_4^{4-} ion (plan)
e.g. as in zircon, $ZrSiO_4$

(a)

$Si_2O_7^{6-}$
e.g. in hemimorphite,
$Zn_4(OH)_2(Si_2O_7).H_2O$

(b)

chain polymer, $(SiO_3)_n^{2n-}$
e.g. as in pyroxenes such as $MgSiO_3$

(c)

O = oxygen   ◉ = silicon

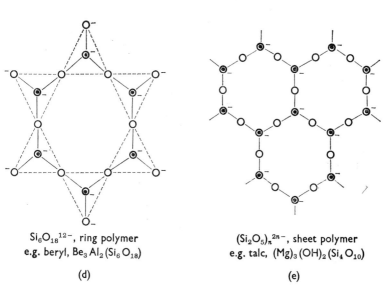

$Si_6O_{18}^{12-}$, ring polymer
e.g. beryl, $Be_3Al_2(Si_6O_{18})$

(d)

$(Si_2O_5)_n^{2n-}$, sheet polymer
e.g. talc, $(Mg)_3(OH)_2(Si_4O_{10})$

(e)

Fig. 8.10

The tendency of silicon to form stable silicon–oxygen chains and networks gives rise to a group of synthetic compounds known as silicones. Hydrolysis of a di-alkyl-dichloro-silane (R_2SiCl_2) gives a chain silicone:

$$
\underset{\overset{\displaystyle |}{R}}{\overset{\overset{\displaystyle R}{|}}{Cl-Si-Cl}} \xrightarrow{H_2O}
\left[\underset{\overset{\displaystyle |}{R}}{\overset{\overset{\displaystyle R}{|}}{HO-Si-OH}} \right]
$$

$$
n\left[\underset{\overset{\displaystyle |}{R}}{\overset{\overset{\displaystyle R}{|}}{HO-Si-OH}} \right] \longrightarrow
\underset{\overset{\displaystyle |}{R}}{\overset{\overset{\displaystyle R}{|}}{-Si}}-O-\underset{\overset{\displaystyle |}{R}}{\overset{\overset{\displaystyle R}{|}}{Si}}-O-\underset{\overset{\displaystyle |}{R}}{\overset{\overset{\displaystyle R}{|}}{Si-}}
$$

chain silicone

Hydrolysis of $RSiCl_3$ gives a cross-linked polymer. The degree of polymerization may be controlled by adjusting the conditions and adding the correct quantity of R_3SiCl, which, having only one active group, terminates the chain-lengthening process. Silicones of various types are used in polishes, lubricants, water-repellents and silicone rubber. A comparison of the properties of carbon and silicon is given in Table 8.5.

Carbon	*Silicon*
(1) Occurs as two allotropic modifications, diamond and graphite.	Only one form—this has the diamond structure.
(2) Readily forms carbon–carbon chains and rings, e.g. the hydrocarbons.	Does not readily form silicon–silicon chains. Never forms silicon–silicon rings.
(3) Does not readily form carbon–oxygen chains and rings.	Readily forms silicon–oxygen chains and rings, e.g. silicates, silicones.
(4) Compounds of the type CX_4 are not very reactive, e.g. to water or air.	Compounds of the type SiX_4 are relatively reactive, e.g. $SiCl_4$ is readily hydrolysed; SiH_4 burns spontaneously in air.
(5) Compounds of the type CX_4 (where X is a halogen) are not lone-pair acceptors (Lewis acids).	Compounds of the type SiX_4 (where X is a halogen) are lone-pair acceptors (Lewis acids).
(6) Forms the stable species $(CN)_2$ and CN^-.	Does not form analogous structures.
(7) Forms a stable monoxide $C\overset{\cdots}{=}O$.	Does not form a stable monoxide.

Table 8.5

Group VM

Sym-bol	Electronic structure	IONIZATION ENERGY (kJ mol⁻¹)		Electro-negativity (Pauling)	RADII (nm)		
		1st	1st + 2nd + 3rd + 4th + 5th		Ionic	Cova-alent	Metallic
N	(He) $2s^2\ 2p^3$	1406	25 698	3·0	0·171 (-3) 0·011 $(+5)$	0·075	—
P	(Ne) $3s^2\ 3p^3$	1063	17 108	2·1	0·212 (-3) 0·034 $(+5)$	0·106	—
As	(Ar) $3d^{10}\ 4s^2\ 4p^3$	967	16 518	2·0	0·222 (-3) 0·047 $(+5)$	0·119	0·139
Sb	(Kr) $4d^{10}\ 5s^2\ 5p^3$	833	14 606	1·9	0·245 (-3) 0·062 $(+5)$	0·138	0·159
Bi	(Xe) $4f^{14}\ 5d^{10}\ 6s^2\ 6p^3$	774	14 722	1·9	0·120 $(+3)$ 0·074 $(+5)$	0·146	0·170

Structural characteristics

(i) All the elements in Group VM have the structure $ns^2\ np^3$ for their valence shells. Arsenic and antimony also have completed d sub-levels in their penultimate shells, while bismuth has a completed d level in its penultimate shell and a completed f level in its ante-penultimate shell.

(ii) The first ionization energies are relatively high and decrease down the group, there being a relatively large drop from nitrogen to phosphorus.

(iii) The electronegativity of nitrogen is high, nitrogen being the third most electronegative element. The electronegativity decreases down the group.

(iv) The $+5$ ionic radii are all small, but increase down the group.

(v) The -3 ionic radii are all relatively large except for that of nitrogen.

(vi) The covalent and metallic radii increase down the group.

(vii) The ionic and covalent radii of nitrogen are much smaller than the corresponding radii of the other elements in the group.

General chemistry

In accordance with the decrease in electronegativity and ionization energies, the metallic character of the Group V elements increases going down the group, so that there is a well-defined gradation of physical properties. Nitrogen is a gas at room temperature and there are no allotropic modifications of the gaseous element. Phosphorus, arsenic and antimony are solids and occur in at least two allotropic forms. One form is metallic in each case and becomes increasingly so going down the group. Bismuth is a solid and has typically metallic characteristics; it occurs only as the metallic form.

The elements of Group V never form ions of the type X^{5+}, because of the small ionic size and high charge of such an ion.

All the elements form tricovalent compounds of the type XH_3 in which the valence shell has the relatively stable octet structure. The more electronegative elements in the group also form ions of the type X^{3-} and so attain an octet structure, e.g. as in Mg_3N_2, magnesium nitride.

The two s electrons in the valence shell show some tendency to remain paired, and this tendency increases going down the group, so that in the common compounds of bismuth the metal has an oxidation number of $+3$; e.g. BiOCl.

When an element in Group V has formed three covalent links, a 'lone pair' of electrons remains. This lone pair may then be donated to a suitable molecule or group to form a coordinate bond; for example,

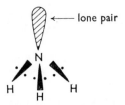

This tendency decreases going down the group.

In the hydrides the lone pair and the three bonding pairs are arranged tetrahedrally so that the molecule is pyramidal; e.g. NH_3,

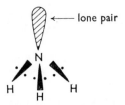

The lone pair, unlike a bonding pair, is associated with only one nucleus, so that it tends to spread out to fill as much space as possible. In so doing it repels the highly localized bonding pairs and so forces the N–H bonds closer together. The H–N–H bond angle is, therefore, less than the theoretical tetrahedral angle (109° 28′). As the electronegativity of the central atom decreases so the lone pair is more 'loosely bound'; thus its repulsive effect on the bonding pairs increases. Consequently the deviation from the theoretical angle increases, as illustrated in Table 8.6.

Ammonia is a weak base since it can donate its lone pair to a proton or other Lewis acid. Phosphine shows some basic properties; e.g. it can form phosphonium iodide, PH_4I, which is analogous to ammonium iodide; but the hydrides of the remaining elements show no basic properties, as the tendency to donate the lone pair decreases down the group.

Hydride	H–X–H bond angle
NH_3	106°45′
PH_3	·94°
AsH_3	91°30′
SbH_3	91°30′
Tetrahedral angle	109°28′

Table 8.6

The elements of Group V form trihalides. These are covalent, except for BiF_3 which is largely ionic, and $BiCl_3$, $BiBr_3$ which show some ionic character. The covalent trihalides are pyramidal like the hydrides. Nitrogen does not form a pentahalide since, being in Period 2, it cannot accommodate more than eight electrons in its valence shell. Phosphorus, arsenic and antimony all form pentahalides, since their valence shells can accommodate more than eight electrons by means of the available d orbitals. No pentahalides of bismuth are known. Phosphorus penta-chloride has a trigonal bipyramid structure as shown in Fig. 8.11. This structure is not as symmetrical as the tetrahedral or octahedral structures, since some bond angles are 90° while others are 120°. In the solid state, PCl_5 tends to revert to an ionic state,

$$2PCl_5 \rightleftharpoons [PCl_4]^+ [PCl_6]^-$$

in which the constituent ions have the highly symmetrical tetrahedral and octahedral shapes.

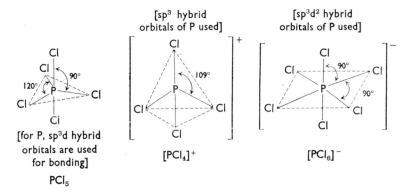

Fig. 8.11

Nitrogen forms a series of stable oxides as shown in Table 8.7. The structures of nitric oxide and nitrogen dioxide are unusual since they each have three electron bonds. The other elements in the group do not form bonds of this type.

Formula	Name	Oxidation number of nitrogen	Structure	Nature
(1) N_2O	Nitrous oxide	+1	Resonance hybrid of $\bar{N}{=}\overset{+}{N}{=}O$ and $N{\equiv}\overset{+}{N}{-}\bar{O}$	Neutral oxide.
(2) NO	Nitric oxide	+2	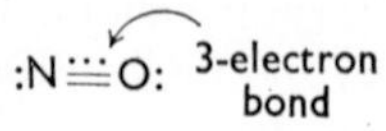 $:N \mathop{::\!:} O:$ 3-electron bond	Neutral oxide.
(3) N_2O_3	Dinitrogen trioxide	+3	Not known	Acidic. Anhydride of nitrous acid HO—N=O.
(4) NO_2	Nitrogen dioxide	+4	Resonance hybrid of 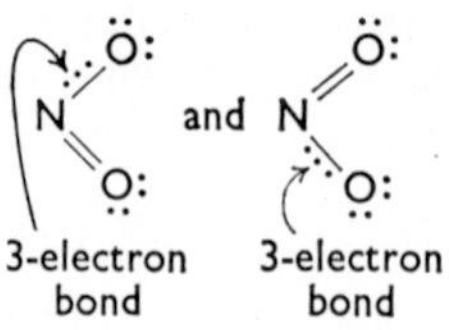 3-electron bond 3-electron bond	Acidic. Mixed anhydride giving HNO and HNO$_3$ with water.
(5) N_2O_4	Dinitrogen tetroxide	+4	Resonance hybrid of 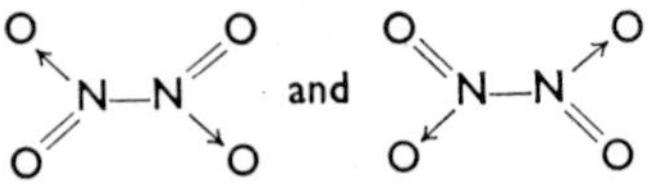	Acidic. Mixed anhydride giving HNO and HNO$_3$ with water.
(6) N_2O_5	Dinitrogen pentoxide	+5	Resonance hybrid of 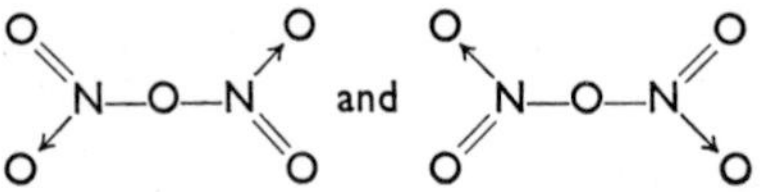 In the solid state: $[NO_2]^+ [NO_3]^-$	Acidic. Anhydride of nitric acid, HNO$_3$.

Table 8.7

Phosphorus forms two oxides of empirical formulae P_2O_3 and P_2O_5. The molecular formulae are P_4O_6 and P_4O_{10}, the structures of both oxides being based on the tetrahedral structure of the P_4 molecule (see

p. 144). This is illustrated in Fig. 8.12. In the P_4O_6 and P_4O_{10} molecules the P–O–P bond angle is approximately 109°. Both the oxides of phosphorus are acidic. Arsenic and antimony form oxides corresponding to those of phosphorus, viz. As_4O_6, Sb_4O_6, Sb_4O_{10}, As_4O_{10}. Their structures are probably the same as the corresponding phosphorus compounds. As_4O_6 is acidic, while Sb_4O_6 is amphoteric. As_4O_{10} dissolves

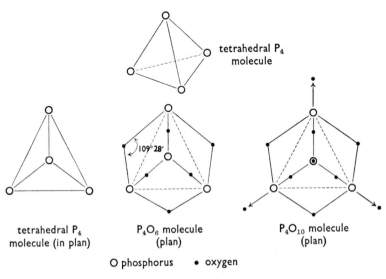

tetrahedral P_4
molecule

tetrahedral P_4
molecule (in plan)

109° 28′

P_4O_6 molecule
(plan)

P_4O_{10} molecule
(plan)

O phosphorus   • oxygen

Fig. 8.12

in water to give arsenic acid, while Sb_4O_{10} is insoluble. Bismuth forms only one oxide, Bi_2O_3, which does not form a dimer and is basic. In general, the basic nature of the trioxides increases down the group as metallic character increases. The stability of the pentoxides decreases down the group as the inert pair effect becomes more important.

Nitrogen forms two important oxyacids, nitric and nitrous:

Nitric acid

Nitrous acid

These give rise to nitrate and nitrite ions in aqueous solution. The nitrate ion has a trigonal structure, and the nitrite ion a 'bent' structure as shown in Fig. 8.13.

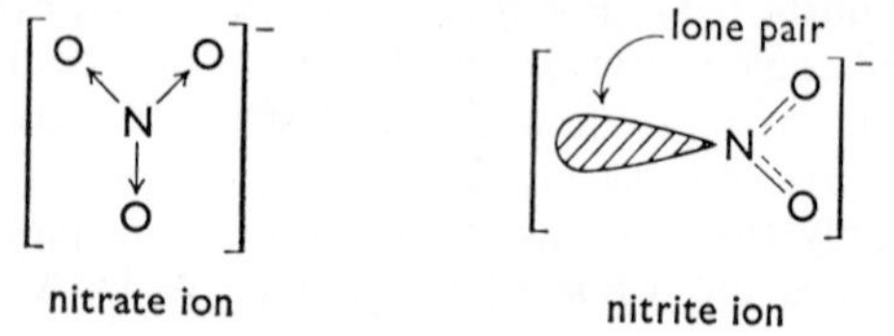

nitrate ion   nitrite ion

Fig. 8.13

Although the vapour of nitric acid probably contains mainly HNO_3 molecules, liquid nitric acid contains a number of species. The particular species present and their concentrations depend upon the dilution of the acid:

$$2H_3O^+ + 2NO_3^- \xrightleftharpoons{+2H_2O} 2HNO_3 \xrightleftharpoons{-H_2O} NO_2^+ + NO_3^-$$

Hydroxonium Nitrate   Nitronium Nitrate
ions   ions   ion   ion

The hydroxonium ions are responsible for the acidic properties of nitric acid, the nitrate ions for its oxidizing properties and the nitronium ions for its action as a nitrating agent. For example, in dilute acid,

$$2H_3O^+ + Mg \longrightarrow Mg^{2+} + 2H_2O + H_2$$

in 50 per cent acid,

$$8H^+ + 2NO_3^- + 3Cu \longrightarrow 3Cu^{2+} + 2NO + 4H_2O$$

and in concentrated acid,

$$C_6H_6 + NO_2^+ \longrightarrow C_6H_5NO_2 + H^+ \text{ (aq)}$$

Phosphorus forms a number of oxy-acids; the structures of some of these, and the ions derived from them, are shown in Tables 8.8 and 8.9. Phosphorus shows some similarity to silicon in its tendency to form polymeric oxy-ions, as illustrated by the polyphosphate ions. The phosphorous acids are reducing agents, the P–H bonds being oxidized to P–OH bonds in redox processes.

Nitric acid is the strongest oxy-acid of the Group V elements, and in general the strengths of analogous oxy-acids decrease going down the group as the electronegativity decreases (see p. 166).

The anomalous properties of nitrogen: comparison with phosphorus

Nitrogen is in Period 2, so that it has a number of anomalous properties. There is a more abrupt change of properties from nitrogen to phosphorus than between any other two consecutive Group V elements.

Name	Structure of acid	Structure of ion
(1) Ortho-phosphoric acid (tribasic)	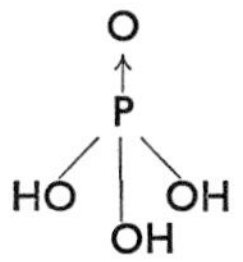 H_3PO_4	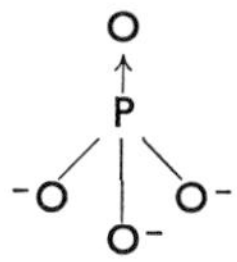
(2) Pyrophos-phoric acid (tetrabasic)	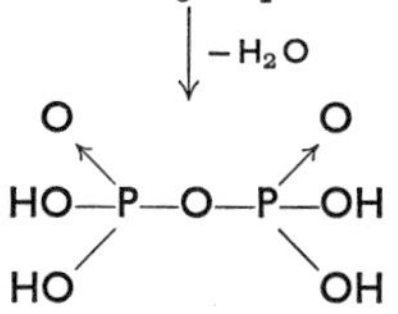 $H_4P_2O_7$	
(3) Tripoly-phosphoric acid (penta-basic)	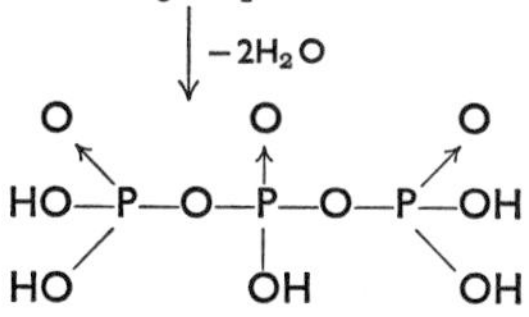	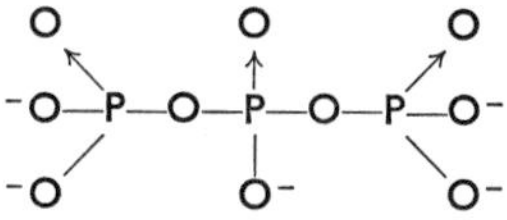

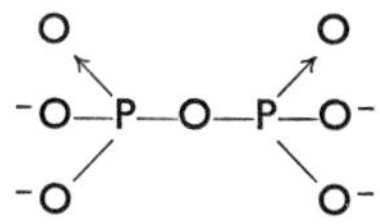

Polymerization continues to give a series of polyphosphoric acids and their corresponding ions.

(4) Meta-phosphoric acid (mono-basic)	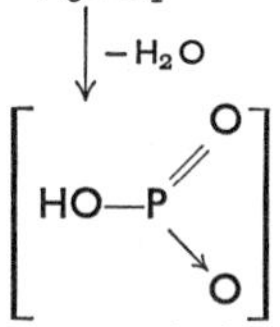 Not formed as such, but polymerizes to give trimetaphosphoric acid.	
(5) Trimeta-phosphoric acid (tribasic)	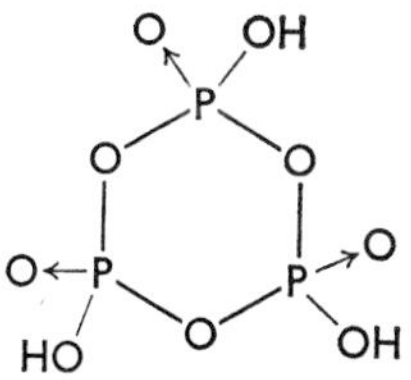 (hypothetical structure)	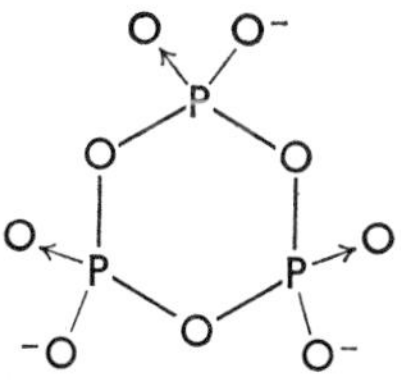

Further polymerization may occur to give tetrametaphosphoric acid and higher ring polymers.

Table 8.8   The phosphoric acid family

Name	Structure of acid	Structure of ion
(1) Ortho-phos-phorous acid (dibasic)	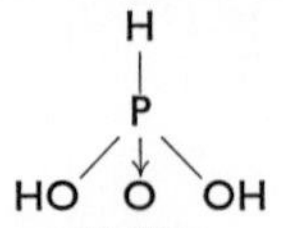 H | P ⟍ ↓ ⟋ HO O OH H_3PO_3	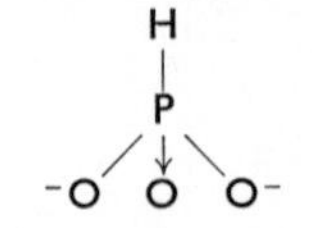 H | P ⟍ ↓ ⟋ ⁻O O O⁻
(2) Pyrophos-phorous acid (dibasic)	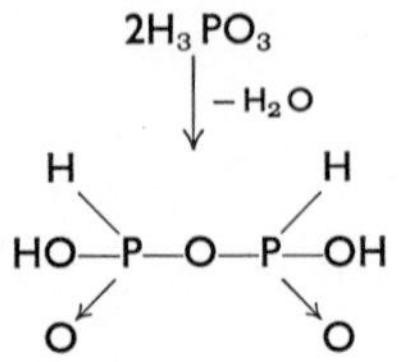 $2H_3PO_3$ ↓ $-H_2O$ H H | | HO—P—O—P—OH ↓ ↓ O O	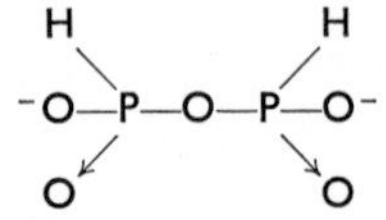 H H | | ⁻O—P—O—P—O⁻ ↓ ↓ O O
(3) Hypo-phos-phorous acid (mono-basic)	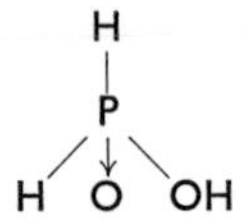 H | P ⟍ ↓ ⟋ H O OH	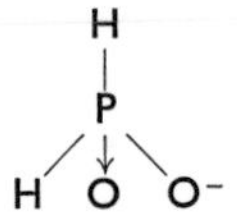 H | P ⟍ ↓ ⟋ H O O⁻

Table 8.9   *The phosphorous acid family*

Nitrogen is much more electronegative than phosphorus. It is one of the four most electronegative elements, the others being fluorine, oxygen and chlorine. The high electronegativity combined with a small atomic radius enables nitrogen to take part in hydrogen bonding; ammonia is much more soluble in water than phosphine because the ammonia molecules form hydrogen bonds with the water molecules:

$$H_3N\text{---}H\text{—}OH$$

The ability of nitrogen to take part in hydrogen bonding is very important in biological systems where proteins are formed from polypeptide chains by hydrogen bond cross-linking:

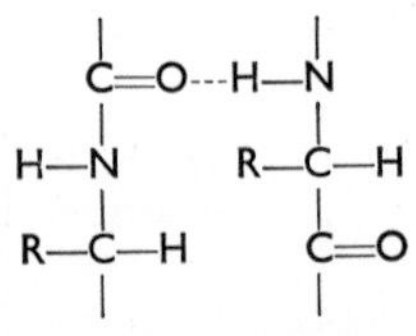

$$
\begin{array}{cc}
\text{C=O---H—N} \\
\text{H—N}\quad\text{R—C—H} \\
\text{R—C—H}\quad\text{C=O}
\end{array}
$$

Nitrogen is saturated when it has eight electrons in its valence shell, while the valence shells of the other elements in the group have d sublevels available and so can hold more than eight electrons. The highest fluoride of nitrogen, therefore, is NF_3, while phosphorus forms PF_5.

Again, nitrogen trichloride is hydrolysed by water to ammonia and hypochlorous acid, while phosphorus trichloride is hydrolysed to orthophosphorous acid and hydrogen chloride. The possible mechanisms for these reactions are:

$$Cl-\underset{\underset{Cl}{|}}{\overset{\overset{Cl}{|}}{N}} + H-OH \longrightarrow Cl-\underset{\underset{Cl}{|}}{\overset{\overset{Cl}{|}}{N}}\cdots H \rightharpoonup OH \longrightarrow Cl-\underset{\underset{Cl}{|}}{\overset{\overset{Cl}{|}}{N}}-H + HOCl$$

in two more similar stages

$$NH_3 + 3HOCl$$

$$Cl-\underset{\underset{Cl}{|}}{\overset{\overset{Cl}{|}}{P}} + :O\underset{H}{\overset{H}{\Big<}} \longrightarrow Cl-\underset{\underset{Cl}{|}}{\overset{\overset{Cl}{|}}{P}}\leftarrow O\underset{H}{\overset{H}{\Big<}} \longrightarrow \underset{Cl}{\overset{Cl}{\Big>}}P-OH + HCl$$

two more similar stages

$$\underset{HO\overset{\downarrow}{\underset{O}{}}OH}{\overset{H}{\underset{|}{P}}} \rightleftharpoons \underset{HO}{\overset{HO}{\Big>}}P-OH$$

Nitrogen trichloride cannot form a dative link with a water molecule because the nitrogen already has eight electrons in its valence shell. The reaction, therefore, occurs by a hydrogen-bonded intermediate. The hydrolysis of phosphorus trichloride proceeds via a dative bond. Because the mechanisms are different the products are not analogous.

Nitrogen forms oxides, e.g. nitric oxide, which contain three-electron bonds. Phosphorus never forms such bonds. Unlike phosphorus, nitrogen forms double and triple bonds with itself, for instance in the nitrogen molecule $N \equiv N$. Gaseous nitrogen does not occur in any allotropic modifications, but phosphorus forms at least two well-defined solid allotropes, white phosphorus and black, as well as a number of modifications of indefinite structure, such as red phosphorus. White phosphorus is made up of P_4 molecules with a tetrahedral structure, as shown in Fig. 8.14a. Black phosphorus is made up of polymeric puckered sheets as shown in Fig. 8.14b. Red phosphorus, although a well-known form, does not appear to have a well-defined structure.

The small interbond angle for white phosphorus (60°) means that there is considerable strain in the molecule, so that white phosphorus is a metastable form and readily reverts to a more stable form where the bond angles are much larger.

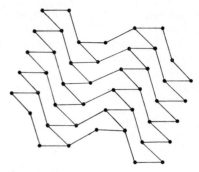

White phosphorus
molecule (P_4)

(a)

Puckered double layer of black phosphorus

(b)

Fig. 8.14

Some important differences between nitrogen and phosphorus are summarized in Table 8.10.

Nitrogen	*Phosphorus*
(1) Gas at room temperature. (b.p. $-195 \cdot 8°C$.)	Solid at room temperature. (b.p. of white phosphorus 280°C.)
(2) Only one modification in the gaseous state.	Forms two well-defined allotropes, white phosphorus and black phosphorus.
(3) The structure of the molecule is $N \equiv N$.	The structure of the molecule in the gaseous state is P $P \diagdown \cdots \diagup P$ P
(4) Is inert to most reagents.	Both allotropes moderately reactive. White form more reactive than black. White phosphorus is spontaneously inflammable in air.
(5) Forms hydrogen bonds.	Does not form hydrogen bonds.
(6) Highest fluoride is NF_3.	Highest fluoride is PF_5.
(7) NCl_3 is hydrolysed to ammonia and hypochlorous acid.	PCl_3 is hydrolysed to ortho-phosphorous acid and hydrogen chloride.
(8) Does not form polymeric oxy-ions.	Forms polymeric oxy-ions, cf. Si.
(9) Forms only two stable acids, HNO_3 and HNO_2. These are stronger than the corresponding oxy-acids of phosphorus.	Forms many oxy-acids, including a series of polymeric acids.
(10) Forms 'odd electron' molecules, e.g. NO.	Does not form 'odd electron' molecules.

Table 8.10

Group VIM

Sym- bol	Electronic structure	IONIZATION ENERGY (kJ mol⁻¹)		Electro- negativity (Pauling)	RADII (nm)		
		1st	1st+2nd +3rd+ 4th+5th +6th		Ionic (calculated)	Cova- lent	Metallic
O	[He] $2s^2\,2p^4$	1314	41 777	3·5	0·140 (−2) 0·009 (+6)	0·073	—
S	[Ne] $3s^2\,3p^4$	1000	26 677	2·5	0·184 (−2) 0·029 (+6)	0·102	—
Se	[Ar] $3d^{10}\,4s^2\,4p^4$	941	25 267	2·4	0·198 (−2) 0·042 (+6)	0·116	0·140
Te	[Kr] $4d^{10}\,5s^2\,5p^4$	870	—	2·1	0·221 (−2) 0·056 (+6)	0·135	0·160
Po	[Xe] $4f^{14}\,5d^{10}\,6s^2\,6p^4$	—	—	2·0	—	—	0·176

Structural characteristics

(i) All the elements in this group have a structure $ns^2\,np^4$ for their valence shells.

(ii) The first ionization energies decrease down the group, though the size of the decrease is much greater from oxygen to sulphur than between any other two consecutive elements.

(iii) The sum of the first six ionization energies shows a general decrease down the group. Again there is a relatively large decrease from oxygen to sulphur.

(iv) The ionic and covalent radii increase down the group.

(v) The electronegativity decreases down the group. The electro- negativity of oxygen is very high, oxygen being the second most electronegative element. (Fluorine, the most electronegative ele- ment, has an electronegativity of 4·0.)

(vi) Oxygen has no d sub-level in its valence shell, which cannot accommodate more than eight electrons.

General chemistry

Oxygen is gaseous, but the other four elements of this group are solids at room temperature. Oxygen has two allotropic modifications. Under normal conditions oxygen is made up of O_2 molecules. These do not have the expected structure,

$$:\!O\!\!=\!\!O\!:$$

since magnetic measurements indicate the presence of two unpaired electrons and in the above structure all the electrons are paired. The O_2 molecule is better represented as

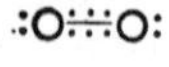

$$:O\vdots\vdots O:$$

The oxygen atoms are joined by one normal σ-bond and two three-electron bonds. If oxygen is subject to a silent electric discharge an allotrope, ozone, is produced. This is a resonance hybrid of the two forms

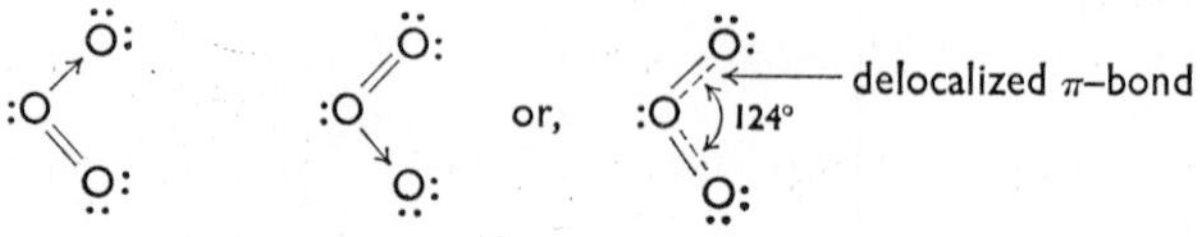

Ozone is metastable and reverts to oxygen when heated. It is a more powerful oxidizing agent than ordinary oxygen.

Sulphur forms a number of well-defined allotropes in the solid state. Two of the better known ones are rhombic (α-) and prismatic (β-) sulphur. Both these forms are made up of S_8 molecules, but the packing in the crystal lattice is different. The S_8 molecule is an eight-membered ring. The bonding pairs and the two lone pairs on each constituent sulphur atom tend to attain a tetrahedral orientation (so as to be as far apart in space as possible); hence the ring is puckered, as shown in Fig. 8.15.

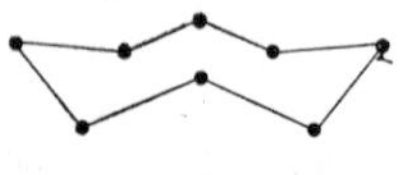

Fig. 8.15

When rhombic sulphur is heated it melts to a mobile liquid made up of randomly moving S_8 molecules. With further heating the liquid becomes increasingly viscous until the viscosity reaches a maximum. Further heating then produces a decrease in viscosity, until at the boiling point (444°C) the liquid is again quite mobile. This change in viscosity may be explained in terms of a change in molecular complexity. As the temperature rises the S_8 rings begin to break up and form long polymeric chains. These cannot slide freely one over the other, but become intertwined so that the viscosity rises. It has been calculated that at the point of maximum viscosity the average chain length is about one million atoms. As the temperature rises still further the chains begin to break up as the vigour of thermal vibration increases, so that the viscosity begins to fall. The sulphur boils at 444°C to give sulphur vapour consisting of S_8 molecules. A further increase in temperature causes a further decrease in molecular complexity to give such species as S_6 and

S_4, until above 900°C the vapour is made up of S_2 molecules. Throughout the heating of liquid sulphur, S_8 rings are always present in equilibrium with the polymeric chains, so that liquid sulphur behaves as a solution of a polymer (sulphur chains) in a monomer (S_8 rings). When liquid sulphur is poured into water, plastic sulphur is obtained. This is made up of randomly arranged S_8 rings and sulphur chains. The arrangement in the liquid has been 'frozen out' by rapid cooling. When plastic sulphur is stretched it crystallizes to give fibrous sulphur. This is made up of a parallel array of sulphur chains with S_8 rings packed regularly between them.

Although sulphur and oxygen are in the same group, oxygen shows little tendency to form chains and rings. This is explained by the small size of the oxygen atom. This means that when two oxygen atoms form a π-bond there is a much greater degree of sideways overlap than for two sulphur atoms. Consequently, the formation of a π- and a σ-bond by oxygen gives a system of lower energy than that obtained by the formation of two σ-bonds. The converse is true for sulphur. Sulphur, therefore, tends to form chains and rings with two σ-bonds per sulphur atom, while oxygen does not.

Selenium atoms also show some tendency to polymerize, and the element forms a number of allotropes containing eight-membered rings and long chains. Two of the allotropes, grey and red crystalline selenium, have the same structural relationship as fibrous and rhombic sulphur.

The elements of Group VI bond to other atoms in three important ways:

(i) *Covalency or electron sharing*

The elements usually share two electrons with other atoms and so attain the relatively stable octet structure; e.g.

The elements sulphur to tellurium also have d levels available in their valence shells, so that they can form more than two covalent bonds by unpairing of paired electrons, promotion to d levels and hybridization; e.g. in sulphur hexafluoride, SF_6. The structure of the sulphur atom in the ground state is:

$$S\ [Ne]\ 3s^2\ 3p_x^2\ 3p_y^1\ 3p_z^1$$

Promotion of a 3s electron and a $3p_x$ electron gives:

	3s	3p			3d			
S [Ne]	1	1	1	1	1	1			

Hybridization then occurs to give six equivalent octahedrally orientated sp^3d^2 orbitals. Overlap with the half-filled p orbitals of fluorine then gives sulphur hexafluoride:

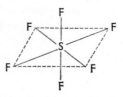

Sulphur, selenium and tellurium form tetrachlorides and tetra-fluorides. Because oxygen is saturated with eight electrons in its valence shell, it always has an oxidation state of 2. Sulphur, selenium and tellurium can have oxidation states of 2, 4 and 6. The oxidation state of 2 arises by the sharing of the two unpaired p electrons and that of 6 by the sharing of electrons in sp^3d^2 hybrid orbitals, as explained above. The formation of tellurium tetrachloride illustrates the bonding and molecular shape for the oxidation state of 4. The process may be represented thus:

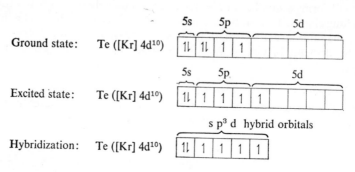

The four half-filled sp^3d hybrid orbitals may then overlap with the half-filled p orbitals of four chlorine atoms to give $TeCl_4$. The electron-pair distribution is trigonal bipyramidal with one of the positions occupied by the 'inert' pair, as shown in Fig. 8.16.

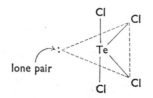

Fig. 8.16

(ii) *Formation of an ion of the type X^{2-}*

When oxygen and sulphur combine with elements of low electronegativity, such as the alkali metals or alkaline earths, they do so by ionic bonding; e.g.

$$Ca^{2+} \quad O^{2-}$$

$$2Na^+ \quad S^{2-}$$

In this case the oxygen (or sulphur) attains a relatively stable octet structure.

(iii) *Formation of a dative bond*

In water the oxygen has eight electrons in its valence shell, but it can combine with a proton by donating a lone pair of electrons, i.e. by the formation of a dative bond:

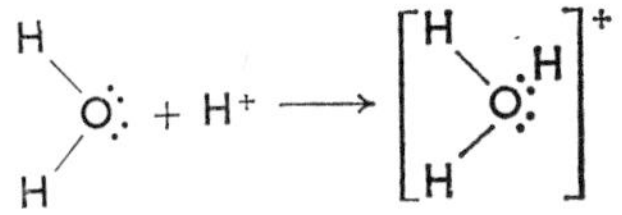

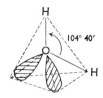

Compounds

Because of the high electronegativity and relatively small size of the oxygen atom, it is able to form hydrogen bonds; e.g. in water:

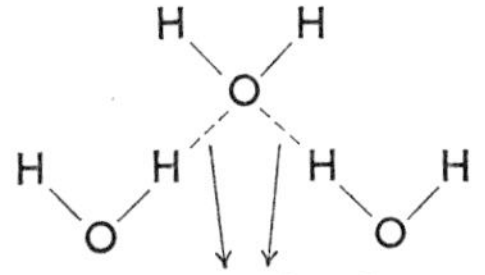

hydrogen bonds

All the elements in Group VI form hydrides of the type H_2X. The hydride molecule is bent, with the lone and bonding pairs occupying tetrahedral positions. For example, the structure of the water molecule is:

104° 40′

The bond angle decreases from H_2O to H_2Te as the electronegativity of the central atom decreases and lone pair, bonding pair repulsion becomes more important. All the hydrides are weakly acidic, the acid strength *increasing* going down the group, as shown in Table 8.11.

Hydride	Equilibrium constant (K_A) for the change $H_2X \rightleftharpoons H^+ + HX^-$
H_2O	$1 \cdot 07 \times 10^{-16}$ (18°C)
H_2S	$9 \cdot 1 \ \times 10^{-8}$ (18°C)
H_2Se	$1 \cdot 7 \ \times 10^{-4}$ (25°C)
H_2Te	$2 \cdot 3 \ \times 10^{-3}$ (25°C)

Table 8.11

This increase is at first sight contrary to expectations, since the electronegativity *decreases* down the group. This would seem to imply that the tendency for H to split off as a proton should decrease down the group as the ionic character of the hydrides decreases. However, the acid strength depends on a number of factors. The decrease in electronegativity down the group is associated with a decrease in bond strength, which favours an increase in acidity. Again for acid dissociation of the type

$$H_2X \rightleftharpoons H^+ + HX^-$$

the size of the anion increases down the group as the ionic radius of X^{2-} increases. But it is a property of electrical systems that the larger the volume over which a given charge is distributed the more stable is the system. Charge dispersal increases down the group as anionic size increases, so that the anion HX^- becomes more stable. This favours increased acidity going down the group. The net result is that the latter two factors outweigh the first, so that acid strength of the hydrides increases down the group. This is summarized below

(1) Stability of H_2X:

$H_2O > H_2S > H_2Se > H_2Te$

tends to *increase* acidity down the group →

(2) Stability of HX^-:
(due to electronegativity of X)

$OH^- > HS^- > HSe^- > HTe^-$

← tends to *decrease* acidity down the group

(3) Stability of HX^-:
(due to ionic size)

$OH^- < HS^- < HSe^- < HTe^-$

tends to *increase* acidity down the group →

(4) Acidity of H_2X:
[as a result of (1), (2) and (3)]

$(H_2O \rightleftharpoons OH^- + H^+) < (H_2S \rightleftharpoons HS^- + H^+) < (H_2Se \rightleftharpoons HSe^- + H^+) < (H_2Te \rightleftharpoons HTe^- + H^+)$

acidity *increases* down the group →

Water has a number of anomalous properties compared to the other hydrides of Group VI. This is illustrated by the boiling points of the hydrides of Groups VI and IV (shown in Fig. 8.17). The boiling points of the Group IV hydrides increase as molecular weight increases; this is the normal trend for any family of compounds. The behaviour of water, however, is anomalous; its boiling point is much higher than those of the other Group VI hydrides. The melting point of water is also anomalously higher. Again, whereas most liquids have their maximum density at their melting points, water has its point of maximum density at $4°C$.

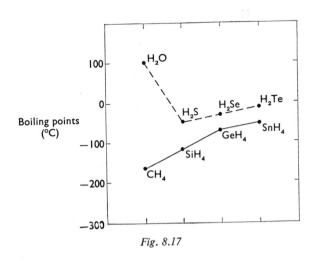

Fig. 8.17

These and other anomalous properties can be explained in terms of the unusual structure of water. To understand this, the structure of ice must first be considered. As already indicated, the water molecule consists of a central oxygen atom with two hydrogen atoms and two lone pairs arranged tetrahedrally about it. Hydrogen bonds can be formed between the hydrogen atoms and the oxygen atom of a neighbouring molecule. These hydrogen bonds will be formed along the direction of maximum electron density, i.e. along the axis of a lone pair. The result is that in a hydrogen-bonded structure each oxygen atom will be surrounded tetrahedrally by four hydrogen atoms, two of which are covalently bonded and two hydrogen-bonded, as shown in Fig. 8.18a. Ice therefore has a tetrahedral network structure similar to that of diamond (Fig. 8.18b). This is confirmed by X-ray diffraction studies. Such a structure is a relatively open one—hence the low density of ice, According to one theory (Bernal and Fowler) liquid water consists of icelike clusters of water molecules held together in tetrahedral arrangements by hydrogen bonds. There are also free molecules of water present and these can move into the hexagonal channels in the hydrogen-bonded clusters. This allows closer packing of the molecules, so that the density increases when ice melts. As the temperature continues to rise two factors must be considered:

(a) the collapse of the open icelike clusters as more hydrogen bonds break,

(b) the increased thermal motion of the molecules which increases the average distance between them.

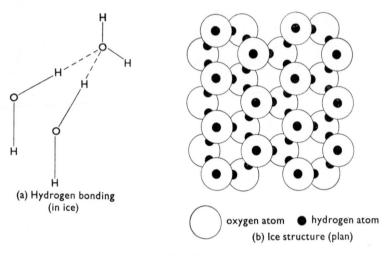

(a) Hydrogen bonding (in ice)

○ oxygen atom   ● hydrogen atom

(b) Ice structure (plan)

Fig. 8.18

Up to 4°C, (a) is the important factor, so the density increases. After 4°C, (b) is the important factor, so the density decreases. If water has the complex hydrogen-bonded structure suggested by the above model it is not surprising that it should have anomalously high boiling and melting points. Pauling has modified the above theory by suggesting that compounds such as $Cl_2.8H_2O$, $Kr.6H_2O$ represent the hydrogen-bonded structures present in liquid water. In these structures (which are called *clathrate* compounds), the water molecules form a hydrogen-bonded cage containing the Cl_2, etc. Pauling's structure is based upon these clathrate cages, with the central molecule (e.g. Cl_2) replaced by a water molecule. This model gives a calculated result for the density of water in good agreement with experiment. Frank has suggested that these hydrogen-bonded groups are dynamic structures, constantly disintegrating and reforming. He calls them 'flickering clusters'. Choppin and Buijs, using infra-red spectroscopy, have estimated that at 40°C the average cluster size is 90 molecules.

The other hydrides of Group VI do not form hydrogen bonds, so the physical properties show the normal change with molecular weight.

Oxygen forms a peroxide, H_2O_2, and sulphur forms a series of poly-sulphides, e.g. H_2S_2, H_2S_3 and H_2S_4, which contain sulphur–sulphur chains such as

$$H—S—S—S—S—H$$

Hydrogen peroxide has the structure shown in Fig. 8.19. It is less stable

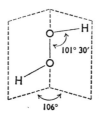

Fig. 8.19

than water and is a strong oxidizing agent. Hydrogen persulphide has a similar structure to hydrogen peroxide. It is less stable than hydrogen sulphide, but is not an oxidizing agent. Hydrogen peroxide may be considered to be the parent acid of the peroxides, which are formed by the more electropositive metals. These contain the peroxide ion, $(O–O)^{2-}$, and give rise to hydrogen peroxide on treatment with dilute acids.

The important oxides of sulphur are the dioxide, SO_2, and the tri-oxide, SO_3. Selenium forms two oxides, SeO_2 and SeO_3, while tellurium forms three, TeO, TeO_2 and TeO_3. Sulphur and selenium dioxide are acidic, giving rise to sulphurous acid, H_2SO_3, and selenious acid, H_2SeO_3, respectively, with water. Tellurium dioxide is amphoteric, dissolving in both acids and alkalis to give salts. This is in accordance with the usual increase in metallic character going down a group. The structure of sulphur dioxide is similar to that of ozone, as shown in Fig. 8.20.

Sulphuric trioxide is acidic, being the anhydride of sulphuric acid.

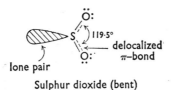

Sulphur dioxide (bent)

Fig. 8.20

The structure of the monomer is shown in Fig. 8.21. At room temperature it polymerizes to give a white solid.

$$\begin{array}{c} :\!\overset{..}{O}: \\ \| \\ S \!\!\swarrow 120° \\ :\!O^{\!\!\nearrow}\; \overset{..}{}\!\! \searrow\! \overset{..}{O}: \end{array}$$

Sulphur trioxide
(trigonal)

Fig. 8.21

Oxygen resembles sulphur more than any other element in the group. With the more electropositive elements they both form ionic compounds, e.g. $Ca^{2+}O^{2-}$, $Ca^{2+}S^{2-}$. They form covalent oxides and sulphides with the more electronegative metals, e.g. SnO_2, SnS_2, and also with non-metals, e.g. CO_2, CS_2. When treated with mineral acids, metallic oxides and sulphides yield a salt and the hydride; for example,

$$Ca^{2+} + O^{2-} + 2H^+ + 2Cl^- \longrightarrow Ca^{2+} + 2Cl^- + H_2O$$
$$Ca^{2+} + S^{2-} + 2H^+ + 2Cl^- \longrightarrow Ca^{2+} + 2Cl^- + H_2S$$

Again, sulphur forms 'thio-' compounds in which a sulphur atom has taken the place of an oxygen atom; e.g. the thiosulphate ion, shown in Fig. 8.22.

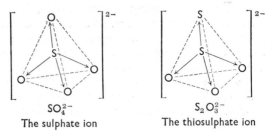

SO_4^{2-}

The sulphate ion

$S_2O_3^{2-}$

The thiosulphate ion

Fig. 8.22

Many oxygen-containing organic compounds have sulphur analogues; for example,

alcohols	R.OH	thiols	R.SH
ethers	R—O—R	thio-ethers	R—S—R

Selenium also forms compounds analogous to alcohols, viz. RSeH.

Sulphur forms a number of oxy-acids, some of which are shown in Table 8.12. Selenium forms only two oxy-acids, selenic acid, H_2SeO_4, and selenious acid, H_2SeO_3.

Name	Structure of acid	Structure of ion
(1) Sulphurous acid (H_2SO_3) (dibasic)	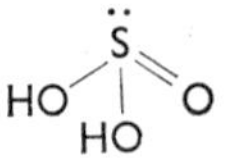	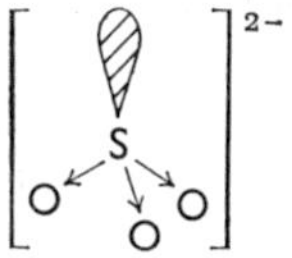 pyramidal
(2) Sulphuric acid (H_2SO_4) (dibasic)	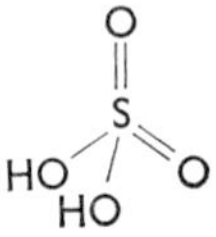	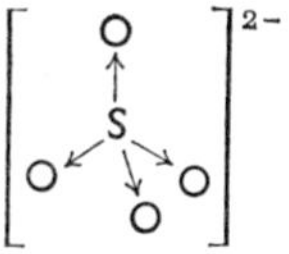 tetrahedral
(3) Thiosulphuric acid ($H_2S_2O_3$) (dibasic)	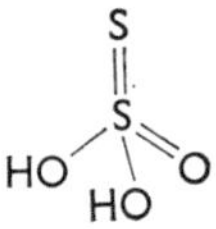	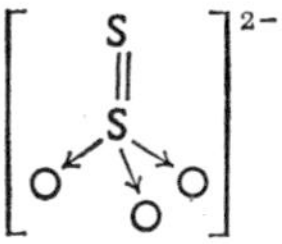
(4) Pyrosulphuric acid ($H_2S_2O_7$) (dibasic)	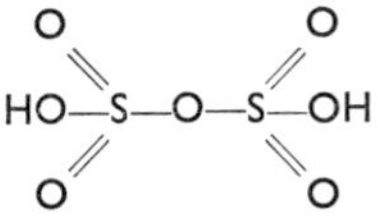	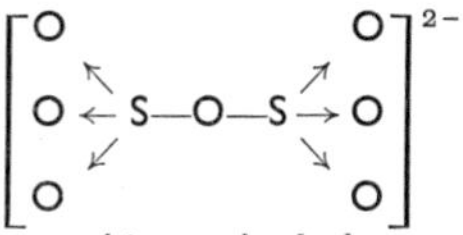 bi-tetrahedral
(5) Dithionic acid ($H_2S_2O_6$) (dibasic)	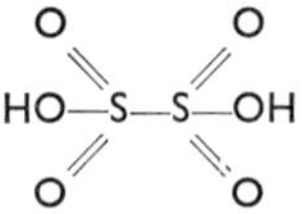	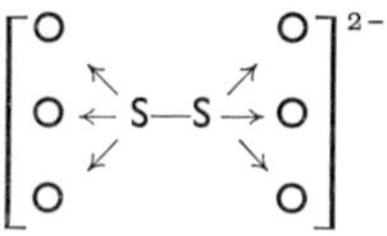
(6) Tetrathionic acid ($H_2S_4O_6$) (dibasic)	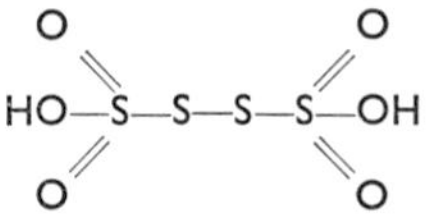	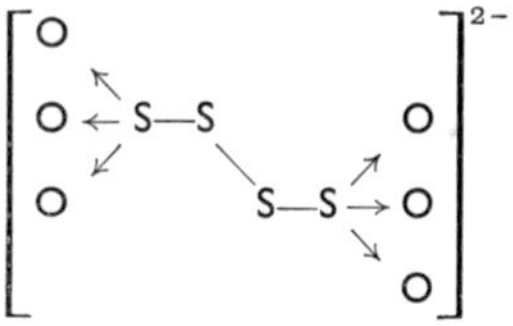
(7) Peroxo-sulphuric acid (H_2SO_5) (dibasic)	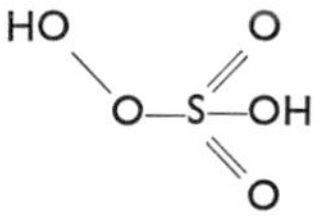	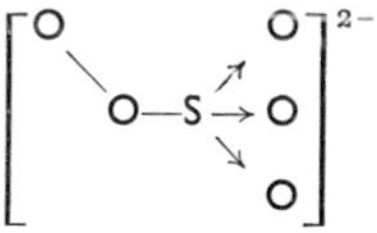
(8) Peroxodi-sulphuric acid ($H_2S_2O_8$) (dibasic)	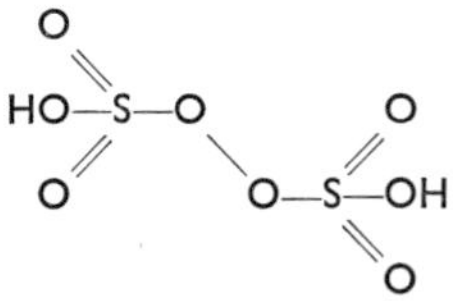	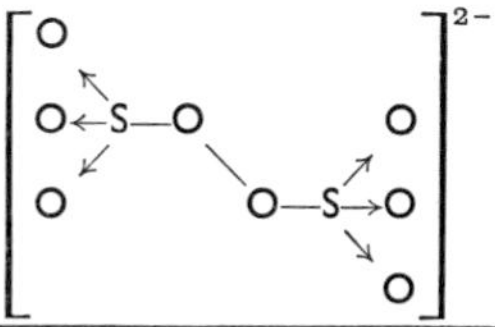

Table 8.12

The Group VI elements form a wide variety of halogen compounds. Oxygen always has an oxidation state of 2 in its halides. Sulphur, selenium and tellurium all have oxidation numbers of 2, 4 and 6, fluorine bringing out the maximum oxidation number of 6. The various halides formed are listed in Table 8.13.

	XY$_2$				XY$_4$				XY$_6$			
	XF$_2$	XCl$_2$	XBr$_2$	XI$_2$	XF$_4$	XCl$_4$	XBr$_4$	XI$_4$	XF$_6$	XCl$_6$	XBr$_6$	XI$_6$
O	OF$_2$ g	OCl$_2$ g	OBr$_2$ dec									
S		SCl$_2$ l			SF$_4$ g	SCl$_4$ l			SF$_6$ g			
Se					SeF$_4$ l	SeCl$_4$ s	SeBr$_4$ s		SeF$_6$ g			
Te		TeCl$_2$ s	TeBr$_2$ s		TeF$_4$ s	TeCl$_4$ s	TeBr$_4$ s	TeI$_4$ s	TeF$_6$ g			

	X$_2$Y$_2$				*Others*
	X$_2$F$_2$	X$_2$Cl$_2$	X$_2$Br$_2$	X$_2$I$_2$	
O	O$_2$F$_2$ g				ClO$_2$, Cl$_2$O$_6$, Cl$_2$O$_7$, BrO$_2$, BrO$_3$, I$_2$O$_4$, I$_4$O$_9$, I$_2$O$_5$
S	S$_2$F$_2$ g	S$_2$Cl$_2$ l	S$_2$Br$_2$ l		S$_2$F$_{10}$
Se	Se$_2$F$_2$ g	Se$_2$Cl$_2$ l	Se$_2$Br$_2$ l		
Te					Te$_2$F$_{10}$

g = gas
l = liquid
s = solid
dec = decomposes
} at room temperature

Table 8.13

The probable molecular shapes of the halides are summarized in Fig. 8.23.

The halogen oxides will be considered in more detail in the next section.

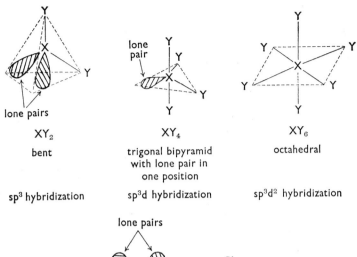

XY₂
bent

sp³ hybridization

XY₄
trigonal bipyramid
with lone pair in
one position

sp³d hybridization

XY₆
octahedral

sp³d² hybridization

X₂Y₂
hydrogen peroxide
structure

sp³ hybridization

Fig. 8.23

Group VIIM, the halogens

Sym-bol	Electronic structure	First ionization energy (kJ mol⁻¹)	Electron affinity (kJ mol⁻¹)	Electro-negativity (Pauling)	RADII (nm)	
					Ionic	Covalent
F	(He) 2s² 2p⁵	1682	347	4·0	0·136 (−1) 0·007 (+7)	0·072
Cl	(Ne) 3s² 3p⁵	1255	368	3·0	0·181 (−1) 0·026 (+7)	0·100
Br	(Ar) 3d¹⁰ 4s² 4p⁵	1142	343	2·8	0·195 (−1) 0·039 (+7)	0·114
I	(Kr) 4d¹⁰ 5s² 5p⁵	1008	314	2·5	0·216 (−1) 0·050 (+7)	0·134
At	(Xe) 4f¹⁴ 5d¹⁰ 6s² 6p⁵	—	—	2·2		—

Structural characteristics

  (i) The Group VII elements have the electronic structure $ns^2 np^5$ for their valence shells.

 (ii) Bromine, iodine and astatine have completed d levels in their penultimate shell. Astatine also has a completed f level in its ante-penultimate shell.

(iii) The first ionization energies are relatively high but decrease down the group. The decrease from fluorine to chlorine is much greater than the decrease between any other two consecutive elements.

 (iv) The electron affinities are high, increasing from fluorine to chlorine and then decreasing down the group.

  (v) The electronegativities are relatively high but decrease down the group.

 (vi) The -1 ionic radii are small and increase down the group. The $+7$ ionic radii are extremely small and increase down the group.

(vii) The covalent radii increase down the group so that the value for iodine is nearly twice that for fluorine.

(viii) Fluorine is an exceptional element in a number of ways:

  (a) It has the highest electronegativity of all the elements.
  (b) It has the second highest electron affinity.
  (c) It has the highest first ionization energy (apart from the noble gases).
  (d) It has the smallest covalent radius (except for hydrogen).
  (e) It has the smallest anion radius.

General chemistry

The halogens are a good example of a group in the Periodic Table, since they show a clear-cut gradation of physical and chemical properties. Some of their physical properties are summarized in Table 8.14.

Element	Physical state	Colour	Melting point (°C)	Boiling point (°C)
F_2	gas	pale yellow	-223	-187
Cl_2	gas	greenish yellow	$-101{\cdot}6$	$-34{\cdot}6$
Br_2	liquid	red-brown	$-7{\cdot}3$	$58{\cdot}7$
I_2	solid	deep violet (lustrous)	$113{\cdot}5$	184

Table 8.14

The halogens are coloured because of the absorption of incident radiation in the visible region of the spectrum. The absorbed radiation excites electrons to higher energy levels. In the small fluorine atom the electrons are 'tightly bound', so only high frequency radiation is absorbed, i.e. light from the blue end of the spectrum; hence fluorine is yellow. As the size of the halogen atom increases radiation of lower frequency is absorbed so that the colours change as indicated until, with iodine, only the extreme violet is not absorbed. As the molecular size increases so the possible 'area of contact' between molecules increases, with a consequent increase in the attractive power of Van der Waals forces. This in turn results in an increase in melting point and boiling point. The lustrous appearance of solid iodine is one sign of increase in metallic character.

In general, metallic character increases down a group. In Group VII fluorine and chlorine show no basic properties. Bromine shows slight basic properties in that bromine trifluoride ionizes:

$$2BrF_3 \rightleftharpoons [BrF_2]^+ + [BrF_4]^-$$

However, there is no evidence for the existence of a simple bromine cation. Iodine shows some distinctly metallic properties; e.g. iodine cyanide conducts electricity when fused and iodine only is liberated at the cathode. This suggests the formation of I^+, thus:

$$ICN \rightleftharpoons I^+ + CN^-$$

Again, electrolysis of fused iodine tri-acetate between silver electrodes yields one mole of silver iodide for every three moles of electrons used. This suggests that iodine tri-acetate ionizes thus:

$$I(OOC.CH_3)_3 \rightleftharpoons I^{3+} + 3CH_3COO^-$$

Since oxidation may be defined as loss of electrons, any substance which is a good electron acceptor will be a strong oxidizing agent. All the halogens, therefore, are oxidizing agents, the oxidizing power decreasing down the group. Since chlorine has the highest electron affinity it might, at first sight, appear that chlorine should be the strongest oxidizing agent. However, the electron affinity corresponds to the energy change for the process

$$X(g) + e^- \longrightarrow X^-(g)$$

while many oxidation processes for the halogens occur in solution and may be represented as follows:

$$\tfrac{1}{2}X_2 + e^- \longrightarrow X^- \text{ (aq)}$$

This may be broken down into a number of hypothetical steps:

$\tfrac{1}{2}X_2(s) \longrightarrow \tfrac{1}{2}X_2(g)$	sublimation (or evaporation if X_2 is a liquid)
$\tfrac{1}{2}X_2(g) \longrightarrow X(g)$	dissociation
$X(g) + e^- \longrightarrow X^-(g)$	electron addition
$X^-(g) \longrightarrow X^-(aq)$	hydration
$\tfrac{1}{2}X_2(s) + e^- \longrightarrow X^-(aq)$	net result

Energy changes for these processes are shown in Table 8.15. When the oxidizing reaction occurs there is an evolution of heat to the surroundings for all the halogens—the heat content of the system decreases.

	Hydration $X^-(g) \rightarrow X^-(aq)$	Electron addition (electron affinity) $X(g) + e^- \rightarrow X^-(g)$	Dissociation (half heat of dissociation) $\frac{1}{2}X_2(g) \rightarrow X(g)$	Sublimation (half heat of sublimation) $\frac{1}{2}X_2(s) \rightarrow \frac{1}{2}X_2(g)$	Oxidizing process (net reaction) $\frac{1}{2}X_2 + e^- \rightarrow X^-(aq)$
F	−514	−347	77·4	—	−783·6
Cl	−372	−368	121	—	−619
Br	−339	−343	96·2	14·6	−571·2
I	−301	−314	75·3	20·9	−518·8

Table 8.15

Since, in general, systems tend to a state of minimum energy and the heat evolved is a maximum for fluorine, then fluorine is the strongest oxidizing agent. The two important energy factors are electron affinity and heat of hydration of the anion. Although the electron affinity of fluorine is lower than that of chlorine the anion has a much larger heat of hydration (because of its small size) so that fluorine is the strongest oxidizing agent. The tendency for the oxidizing power of the halogens to decrease down the group is further confirmed by the values of their standard electrode potentials:

$$F_2 + 2e^- \longrightarrow 2F^- \qquad E^0 = -2\cdot87 \text{ volts}$$

$$Cl_2 + 2e^- \longrightarrow 2Cl^- \qquad E^0 = -1\cdot36 \text{ volts}$$

$$Br_2 + 2e^- \longrightarrow 2Br^- \qquad E^0 = -1\cdot07 \text{ volts}$$

$$I_2 + 2e^- \longrightarrow 2I^- \qquad E^0 = -0\cdot53 \text{ volts}$$

The decrease in oxidizing power is illustrated by the reactions of the halogens with water. Fluorine oxidizes water to oxygen:

$$F_2 + H_2O \longrightarrow 2H^+ + 2F^- + \tfrac{1}{2}O_2$$

while chlorine and bromine react to give a halogen and an oxyacid:

$$Cl_2 + H_2O \longrightarrow HCl + HClO$$

$$Br_2 + H_2O \longrightarrow HBr + HBrO$$

In the case of iodine the reverse process can occur—oxygen oxidizes iodide ions to iodine:

$$2H^+ + 2I^- + \tfrac{1}{2}O_2 \rightleftharpoons I_2 + H_2O$$

The reaction of fluorine, chlorine and bromine with cold dilute alkalis follows a similar pattern:

$$F_2 + 2OH^- \longrightarrow 2F^- + H_2O + \tfrac{1}{2}O_2$$

$$Cl_2 + 2OH^- \longrightarrow Cl^- + ClO^- + H_2O$$

$$Br_2 + 2OH^- \longrightarrow Br^- + BrO^- + H_2O$$

The last two reactions are interesting since the halogens undergo disproportionation. Consider chlorine for example. The oxidation number of Cl in Cl_2 is 0, in Cl^- it is -1, and in ClO^- it is $+1$, so that

$$2Cl\,(0) \quad \text{gives} \quad Cl\,(-1) \quad \text{and} \quad Cl\,(+1)$$

which is disproportionation, i.e. one halogen atom is oxidized at the expense of another, which is reduced.

The halogens are, in general, very reactive. They react with both metals and non-metals. Fluorine is the most reactive because of its high electronegativity, high electron affinity, small atomic and anionic size and low dissociation energy. When fluorine reacts with other elements it tends to bring out their highest oxidation number; e.g. PF_5, SF_6, IF_7. In contrast, the highest chlorides of these elements are PCl_5, SCl_4 and ICl_3.

The halogens form several different types of compound.

(*a*) *Covalent compounds* A halogen atom has seven electrons in its valence shell, so it can attain the relatively stable octet configuration by forming a single covalent bond with another atom. All the halogens form compounds in this way; e.g. H–F, H–Cl, H–Br, H–I. Fluorine can only form one covalent bond since its valence shell can contain a maximum of eight electrons, but the other halogens have d levels available and so can form more than one covalent bond. Consider for example the formation of IF_7. This may be represented as follows:

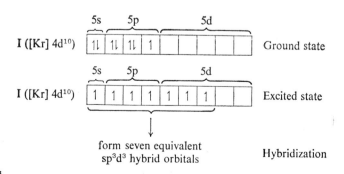

Overlap then occurs between these hybrid orbitals and the half-filled p orbitals of fluorine to give IF_7, as shown below:

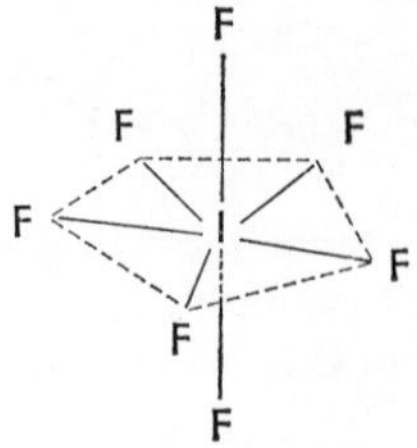

IF_7, pentagonal bipyramid

(*b*) *Ionic compounds*   When the halogens combine with the more electropositive elements they gain an electron to give a halide ion of the type X^-. In so doing they attain the relatively stable octet structure; for example,

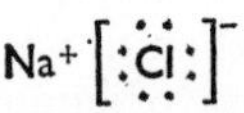

$$Na^+ \left[\ddot{\underset{\cdot\cdot}{Cl}} \right]^-$$

(*c*) *Complex ions*   The halogens form a wide variety of complex ions in which a halide ion ligates a central metal ion; for example,

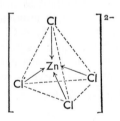

(*d*) *Polyhalide ions*   Halide ions can associate with halogen molecules to give polyhalide ions. This usually occurs in association with a large cation such as an alkali metal or alkaline earth cation; examples are

$$K^+[I—I—I]^- \quad and \quad Cs^+[F—I—Br]^-$$

(*e*) *Hydrogen bonding*   Because of its very high electronegativity and small atomic radius, fluorine forms hydrogen bonds quite readily. This gives rise to the anomalous physical properties of hydrogen fluoride discussed below.

The halogens form hydrides of the type HX. Some of their physical and chemical characteristics are shown in Table 8.16.

The boiling points show a similar pattern to those of the Group VI

Property	HF	HCl	HBr	HI
Boiling point (°C)	19·54	−94·9	−66·8	−35·4
Heat of formation at 20°C (kJ mol⁻¹)	269	91·6	30·5	5·52
Apparent degree of dissociation in				
0·1M soln. (18°C)	0·10	0·926	0·93	0·95
Internuclear distance (nm)	0·092	0·1276	0·1410	0·162

Table 8.16

hydrides. The volatility decreases from hydrogen chloride to hydrogen iodide in the expected manner. Hydrogen fluoride, however, is much less volatile than the other hydrides, being a liquid at room temperature; it has an anomalously high boiling point because it is highly associated, hydrogen bonding giving rise to zig-zag chains of HF molecules:

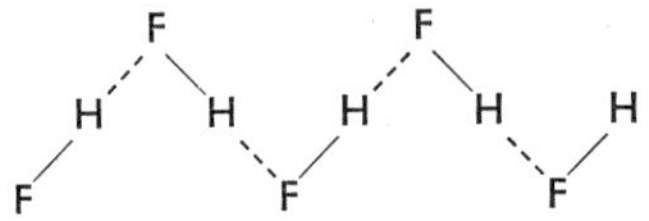

The hydrogen bonding is strong enough to give rise to 'acid' salts of hydrogen fluoride, such as $K^+[HF_2]^-$ and $K^+[H_2F_3]^-$.

The degrees of dissociation shown in Table 8.16 show an increase from HF to HI, i.e. acid strength increases down the group. This behaviour is similar to that of the Group VI hydrides and can be explained along the same lines. The decrease in electronegativity down the group tends to decrease the readiness with which the X^- ion is formed, and so decrease the acidity down the group. However, the decrease in electronegativity is also associated with a decrease in bond strength (see heats of formation, Table 8.16) which tends to increase acidity down the group. The increase in anion size down the group tends to stabilize X^-, which results in increased acidity down the group. The net result is that the last two factors are the more important so that acidity increases down the group.

Fluorine can only exert a covalency of one, and so forms only two oxides

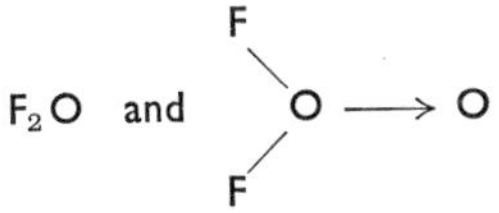

The other halogens, however, form a number of oxides. Table 8.17 lists these and gives their structures, where known.

Fluorine	Chlorine	Bromine	Iodine
(1) Oxygen difluoride OF_2  Oxidation number of F = −1	(1) Dichlorine monoxide Cl_2O 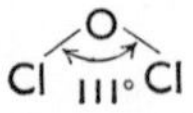 Oxidation number of Cl = +1	(1) Bromine monoxide Br_2O 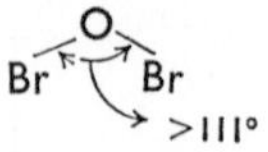 Oxidation number of Br = +1	(1) Di-iodine pentoxide I_2O_5 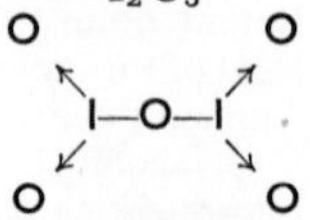 (probable) Oxidation number of I = +5
(2) Di-oxygen difluoride O_2F_2 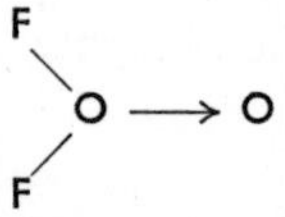 (possible) Oxidation number of F = −1	(2) Chlorine dioxide ClO_2 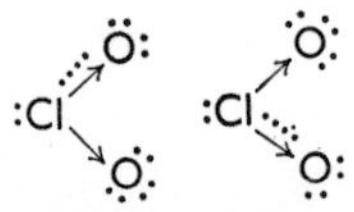 resonance hybrid of above structures Oxidation number of Cl = +4	(2) Bromine dioxide BrO_2 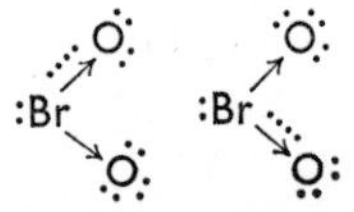 resonance hybrid of above structures Oxidation number of Br = +4	(2) I_2O_4 and I_4O_9 Both of these compounds are ionic. They are probably iodates rather than true oxides. Their probable structures are: $IO^+IO_3^-$ $I^{3+}(IO_3^-)_3$ I_2O_4 I_4O_9
	(3) Dichlorine hexoxide Cl_2O_6 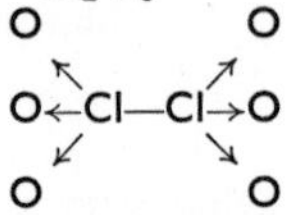 (probable) Oxidation number of Cl = +6	(3) Bromine trioxide BrO_3 Structure unknown Oxidation number of Br = +6	
	(4) Dichlorine heptoxide Cl_2O_7 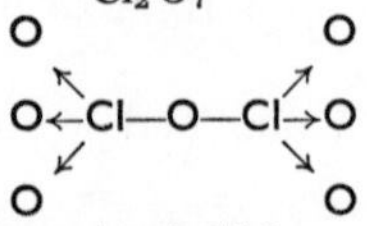 (probable) Oxidation number of Cl = +7		

Table 8.17

The interbond angles increase from F_2O to Br_2O as the size of the halogen atoms increase.

Fluorine forms no oxy-acids, but the other halogens form all or some of the series HXO, HXO_2, HXO_3, HXO_4. Their structures are given in Table 8.18.

	Chlorine	*Bromine*	*Iodine*
HOX	Hypochlorous acid 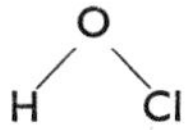 $K_a\ 3\cdot7\times10^{-8}$	Hypobromous acid 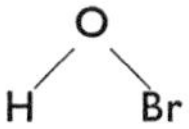 $K_a=2\times10^{-9}$	Hypoiodous acid 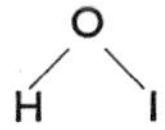 $K_a=1\times10^{-11}$
HXO_2	Chlorous acid 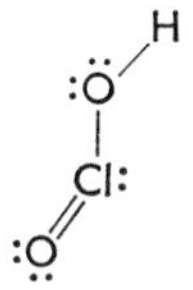	—	—
HXO_3	Chloric acid 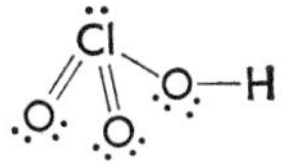 K_a is high	Bromic acid 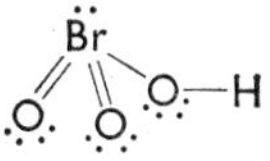	Iodic acid 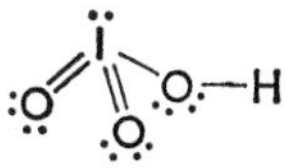 $K_a=1\cdot9\times10^{-1}$
HXO_4	Perchloric acid 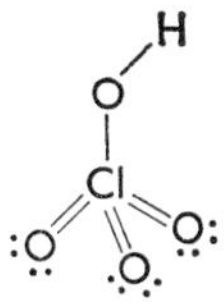 K_a is very high	—	Periodic acid 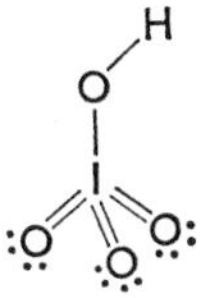 K_a for $HIO_4\,(H_2O)$ $=1\times10^{-3}$

Table 8.18

The oxyacids of the halogens show a regular change in acid strength, (*a*) going down the group for a given acid type, and (*b*) as the number of oxygens present increases for a given halogen. Acid strength decreases in the series

$$HOCl > HOBr > HOI$$

but increases in the series

$$HOCl < HClO_2 < HClO_3 < HClO_4$$

Acidity depends upon the tendency for the hydrogen in the hydroxyl group of the acid to split off as a proton. This tendency will be increased if the electron density near the hydrogen is decreased by any factor. For any oxy-acid

$$X—O—H$$

the greater the electronegativity of X, the greater its electron-withdrawing power and so the greater the tendency for H to split off as a proton. This means that as the electronegativity of X increases so the acid strength of HOX increases. Hence the hypohalous acids decrease in strength down the group. If an additional oxygen is bonded to X then,

Type	Compounds formed	Structure	
XY	ClF (g) BrF (g) BrCl (g) ICl {α (s) β (l) IBr (s)	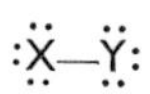 :Ẍ—Ÿ:	linear
XY₃	ClF₃ (g) BrF₃ (l) ICl₃ (s)	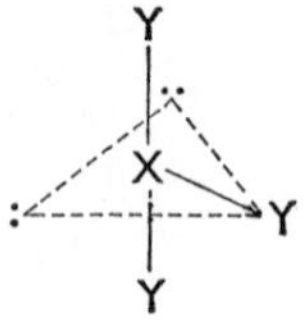	T-shaped
XY₅	BrF₅ (l) IF₅ (l)	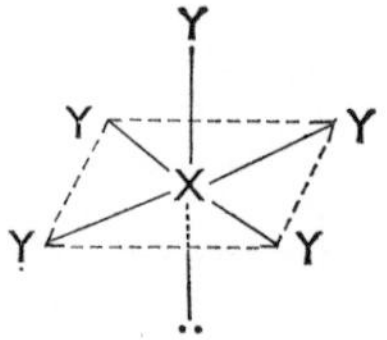	square-based pyramid
XY₇	IF₇ (g)	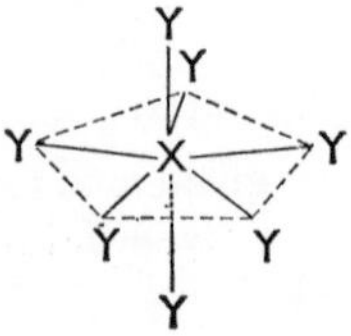	pentagonal bi-pyramid

g = gas
l = liquid } at room temperature
s = solid

Table 8.19

being highly electronegative, it makes X more positive and so increases its electron-withdrawing power so that the acid strength increases. The greater the number of oxygen atoms bonded to the central atom, the greater this effect and so the greater the acid strength. Therefore, acid strength increases in the series hypochlorous acid, chlorous acid, chloric acid, perchloric acid.

Hypochlorous acid is the most stable of the hypohalous acids, which are known only in solution. They are all strong oxidizing agents. Chlorous acid exists only in solution. Chloric and bromic acids are known only in solution but their salts are quite stable. Iodic acid exists as a white solid. Perchloric acid can be prepared in the anhydrous state as a covalent liquid. The common form of periodic acid is the hydrate, $HIO_4.2H_2O$. All the oxyacids are strong oxidizing agents.

The halogens combine directly with each other under the appropriate conditions to form a number of interhalogen compounds. The formulae and structures of these are shown in Table 8.19.

Bromine trifluoride is of practical interest since it is a good, but not too vigorous fluorinating agent. It has been used as a non-aqueous solvent since it shows considerable self-ionization:

$$2BrF_3 \rightleftharpoons BrF_2^+ + BrF_4^-$$

The anomalous properties of fluorine

Fluorine, like the other elements in Period 2, shows many properties which are not typical of the other elements in its group. This is because (a) with eight electrons in its valence shell a main energy level is completed, i.e. it is saturated when it has formed one covalent or ionic bond, (b) it has a much higher electronegativity and first ionization energy than the other elements in the group, and (c) it has much smaller ionic and covalent radii than the other members of the group. Some of the properties which are associated with these atypical structural characteristics are as follows.

(1) It always has an oxidation number of -1.
(2) It forms only F_2O and F_2O_2, no higher oxides.
(3) It does not form any oxy-acids or oxy-acid salts.
(4) Its hydride has an anomalously high boiling point and melting point because of hydrogen bonding.
(5) Hydrogen bonding leads to the formation of the HF_2^- ion.
(6) Fluorine oxidizes water to oxygen.
(7) It tends to bring out the highest oxidation state of other elements when it combines with them; e.g. IF_7.
(8) Its bonds with other elements have the maximum percentage ionic character.

Halogenoids or pseudo-halogens

There are a number of substances which possess properties comparable with those of the halogens. Such substances are called halogenoids or pseudo-halogens. Two examples of these are cyanogen, $(CN)_2$, and thiocyanogen, $(SCN)_2$. Some of the important ways in which they resemble the halogens are as follows.

(1) They combine directly with many metals to give salts; for example,

$$(CN)_2 + 2Ag \longrightarrow 2AgCN$$

(2) The silver, lead (II) and mercury (I) salts so formed are insoluble in water.

(3) They form compounds which are similar in composition and properties to those of the halogens; for example,

$$ICN \qquad \text{compared with} \quad ICl$$

$$Si(SCN)_4 \quad \text{compared with} \quad SiCl_4$$

(4) They often react with hydroxyl ions in a similar way to halogens; for example,

$$(CN)_2 + 2OH^- \longrightarrow CN^- + CNO^- + H_2O$$

$$Cl_2 + 2OH^- \longrightarrow Cl^- + ClO^- + H_2O$$

(5) Their hydracids often react with oxidizing agents in a similar way to the halogen hydracids; for example,

$$MnO_2 + 4H^+ + 2Cl^- \longrightarrow Mn^{2+} + 2H_2O + Cl_2$$

$$MnO_2 + 4H^+ + 2SCN^- \longrightarrow Mn^{2+} + 2H_2O + (SCN)_2$$

(6) Their lead (IV) compounds are decomposed by heat:

$$Pb(SCN)_4 \xrightarrow{\text{heat}} Pb(SCN)_2 + (SCN)_2$$

$$PbCl_4 \xrightarrow{\text{heat}} PbCl_2 + Cl_2$$

(7) They often form addition compounds with alkenes,

$$H_2C{=}CH_2 + Cl_2 \longrightarrow \underset{\underset{Cl}{|}}{\overset{\overset{H}{|}}{H{-}C}}{-}\underset{\underset{Cl}{|}}{\overset{\overset{H}{|}}{C{-}H}}$$

$$H_2C{=}CH_2 + (SCN)_2 \longrightarrow \underset{\underset{SCN}{|}}{\overset{\overset{H}{|}}{H{-}C}}{-}\underset{\underset{SCN}{|}}{\overset{\overset{H}{|}}{C{-}H}}$$

Group VIIIM, the noble gases

Symbol	Electronic structure	First ionization energy (kJ mol^{-1})	Van der Waals radius (nm)
He	$1s^2$	2372	—
Ne	(He) $2s^2\ 2p^6$	2079	0·16
Ar	(Ne) $3s^2\ 3p^6$	1519	0·191
Kr	(Ar) $3d^{10}\ 4s^2\ 4p^6$	1351	0·200
Xe	(Kr) $4d^{10}\ 5s^2\ 5p^6$	1172	0·22
Rn	(Xe) $4f^{14}\ 5d^{10}\ 6s^2\ 6p^6$	1038	—

Structural characteristics

(i) All sub-levels are complete, so that all electrons are paired.
(ii) The first ionization energies are very high but decrease down the group.
(iii) The Van der Waals radii increase down the group.

General chemistry

All the noble gases except radon are present in small quantities in the atmosphere and can be obtained by the fractionation of liquid air. Helium is also found in certain natural gases.

Ever since their discovery (1894–1900) the noble gases have been characterized chiefly by their resistance to attack by chemical reagents. This lack of reactivity fitted in well with the development of valence theory. The lack of reactivity arises because the noble gases possess no unpaired electrons and no incomplete sub-shells. At one stage the tendency to attain a noble gas structure was greatly emphasized in explaining compound formation. More recently, however, well-defined compounds of certain of the noble gases have been prepared, so their structures are now regarded as only *relatively* stable. The important criterion for bond formation is electron pairing, but if this is combined with the closing of an electron shell, i.e. the attainment of a noble gas or octet structure, a relatively very stable compound often results.

Because of their relatively stable electronic configurations the Group VIII elements are monatomic gases. Their physical properties show an almost ideal variation with atomic number. Some of these properties are summarized in Table 8.20.

The melting points and boiling points are very low compared to other materials of comparable molecular weights. This is because the only intermolecular forces are very weak Van der Waals attractions. The specific heat ratios clearly indicate monatomic gases. Those solid phases

Property	He	Ne	Ar	Kr	Xe	Rn
Atomic number	2	10	18	36	54	86
Melting point (K)	0·9(?)	24·43	83·9	104	133	202
Boiling point (K)	4·216	27·2	87·4	121·3	163·9	211·3
Ratio of molar heats (c_p/c_v)	1·65	1·64	1·65	1·69	1·67	—
Water solubility (cm^3dm^{-3} at 20°C)	13·8	14·7	37·9	73	110·9	—

Table 8.20

which have been examined have close-packed cubic structures. The solubilities of the gases in water are quite high, especially for those gases of higher atomic number. A solvent of high permanent dipole, such as water, probably induces a dipole in the noble gas atoms and so increases attractive forces. This effect would increase as the size of the noble gas atom increases, which probably accounts for the increase in solubility of the noble gases as the atomic number increases.

Liquid helium is of interest since it occurs in two forms, helium I and helium II. Helium I is a normal liquid, but helium II has very unusual properties. It is a liquid with the properties of a gas—a superfluid. Its energy is so low that thermal motion of the atoms has almost ceased but the interatomic forces are not strong enough for a solid to be formed. Its viscosity is only one hundredth that of gaseous hydrogen. Its thermal conductivity is eight hundred times as great as that of copper and it can flow in a superfluid film up the sides of the containing vessel. Helium II probably has an open structure, having many atoms in low energy states, among which a few atoms in higher energy states move.

Compounds of the noble gases

(a) Compounds formed under excited conditions
Under conditions of electric discharge or electron bombardment sufficient energy is available for electrons in a noble gas atom to be excited to higher energy levels so that compound formation becomes possible. Consider, for example, the helium atom. In a discharge tube this may undergo the following change:

$$\text{He } 1s^2\, 2s^0 \xrightarrow{\text{energy}} \text{He } 1s^1\, 2s^1$$
$$\text{ground state} \qquad\qquad \text{excited state}$$

Compounds such as He_2^+ and HeH^+ have been observed under these conditions, but such species have only a transitory existence.

(b) Compounds formed by induced dipole–dipole interaction
The nature of this type of bonding has already been discussed above when the water solubility of the noble gases was explained. This type of

interaction probably accounts for the formation of certain phenol derivatives, such as $Kr (C_6 H_5 OH)_2$.

(c) Clathrate or 'cage' compounds

The hydrates of the noble gases and the compounds formed with hydroquinone have been shown to be clathrate compounds. These are inclusion compounds in which noble gas atoms are trapped in a cage of water molecules or hydroquinone molecules. An example is shown in Fig. 8.24.

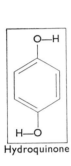

Hydroquinone

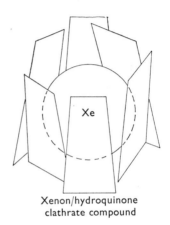

Xenon/hydroquinone
clathrate compound

Fig. 8.24

Helium and neon do not form such compounds because their small size prevents their being trapped in the molecular 'cage'. There are no true chemical bonds in these compounds, only weak Van der Waals forces.

(d) Compounds containing 'true' chemical bonds

During his investigations of the properties of the gas platinum hexafluoride ($Pt^{VI}F_6$), in 1962, N. Bartlett found that it combined with molecular oxygen to give a compound $O_2^+ Pt^V F_6^-$. Barlett realized that the ionization energy of molecular oxygen was very similar to that of xenon:

$$O_2 \longrightarrow O_2^+ + e^- \quad 1180 \ kJ \ mol^{-1}$$
$$Xe \longrightarrow Xe^+ + e^- \quad 1172 \ kJ \ mol^{-1}$$

This suggested that xenon might also combine with platinum hexafluoride. This proved to be the case, and Bartlett prepared the first 'true' compound of the noble gases, $XePtF_6$. This was followed by the preparation of $Xe(RuF_6)_2$ by reacting xenon with ruthenium hexafluoride. Ruthenium and platinum hexafluoride are decomposed by heat:

$$RuF_6 \longrightarrow RuF_5 + \tfrac{1}{2}F_2$$
$$PtF_6 \longrightarrow PtF_4 + F_2$$

and since one atom of xenon reacts with two molecules of RuF_6 but with only one molecule of PtF_6, this suggests that the hexafluorides might be fluorinating the xenon. This led to the investigation of the reaction of xenon and fluorine directly. Fluorine and xenon in the ratio of 5 to 1 by volume were heated to $400°C$ in a nickel can for one hour. The can was cooled to $78°C$ and excess gas removed. Colourless crystals of a xenon fluoride were found in the can. These turned out to be XeF_4. If the conditions of fluorination are varied XeF_2 and XeF_6 can be prepared. When xenon hexafluoride is partially hydrolysed, $XeOF_4$ is obtained:

$$XeF_6 + H_2O \longrightarrow XeOF_4 + 2HF$$

Complete hydrolysis in acid solution yields a solution of xenon trioxide:

$$XeF_6 + 3H_2O \longrightarrow XeO_3 + 6HF$$

Hydrolysis of the hexafluoride in basic solution yields a salt sodium perxenate, $Na_4XeO_6.yH_2O$, in which the xenon has an oxidation number of $+8$. The physical properties of the xenon compounds mentioned are summarized in Table 8.21.

Property	XeF_2	XeF_4	XeF_6	$XeOF_4$	XeO_3	Na_4XeO_6
Melting point (°C)	140	114 (approx)	46	-28	Explodes	decomposes above 160°C
Colour of solid	white	white	white < 42°C yellow > 42°C	white	white	white
Colour of vapour	colourless	colourless	yellow	colourless	—	—

Table 8.21

The xenon compounds do not contain any peculiar types of bonds and current valence theory can account quite readily for the nature of the bonding and the structures of the compounds. To account for their structures the electron pair repulsion theory has been most successful. This assumes:

(i) Bonding and non-bonding pairs tend to get as far away from each other as possible.

(ii) Non-bonding pairs (lone pairs) occupy more space than bonding pairs and so tend to get as far away from each other as possible.

(iii) The two bonding pairs in double bonds behave like a single bonding pair but occupy more space, so these tend to get as far away from each other and from lone pairs as possible.

Consider, for example, the structure of xenon difluoride. The electronic structure of the outer shells of xenon in the ground state may be represented as follows:

Xe ([Kr] $4d^{10}$)

Promotion of an electron followed by hybridization gives

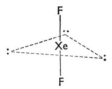

Xe ([Kr] $4d^{10}$)

hybridization to give
five equivalent sp^3d orbitals

Overlap may then occur with the unpaired p electron on two fluorine atoms to give XeF_2. The electron pair repulsion theory predicts the following structure:

i.e. XeF_2 is a linear molecule. This is confirmed by physical measurements.

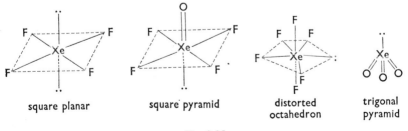

square planar   square pyramid   distorted   trigonal
                                                octahedron   pyramid

Fig. 8.25

The structures of the other xenon compounds are given in Fig. 8.25. These have been confirmed by physical measurements.

References

Books

Concise Inorganic Chemistry, J. D. Lee (Van Nostrand)
Inorganic Chemistry: An Advanced Textbook, T. Moeller (Wiley)
Advanced Inorganic Chemistry, F. A. Cotton and G. Wilkinson (Wiley)
Comparative Inorganic Chemistry, B. J. Moody (Arnold)
Structural Principles in Inorganic Compounds, W. E. Addison (Longmans)
The Allotropy of the Elements, W. E. Addison (Oldbourne)
College Chemistry, L. Pauling (Freeman)

Papers

'Water', A. M. Buswell and W. H. Rodebush (reprint from *Scientific American*, April 1956)
'Water, H_2O or $H_{180}O_{90}$?', G. R. Chopping (*Chemistry*, **38**, No. 3, March 1965, pp. 7–11)
'The Noble Gas Compounds', C. L. Chernick (reprint from *Chemistry*, No. 10, January 1964)
'Structures of the Noble Gas Compounds' (*Chemistry*, **39**, No. 4 (1966), pp. 17–19)

Films

A Research Problem: Inert (?) Gas Compounds, G. C. Pimentel and J. J. Turner, a CHEM Study Film, available from Sound-Services Ltd., cat. no. 4160/999

9. The d-block elements

Introduction

The electronic structure of calcium is

$$\text{Ca } 1s^2 \; 2s^2 \; 2p^6 \; 3s^2 \; 3p^6 \;\vdots\; \boxed{3d^0} \;\vdots\; 4s^2$$

According to the energy diagram given on p. 120 the next lowest energy level is not the 4p but the 3d, so that the next element, scandium, has the structure

$$\text{Sc } 1s^2 \; 2s^2 \; 2p^6 \; 3s^2 \; 3p^6 \;\vdots\; \boxed{3d^1} \;\vdots\; 4s^2$$

The 3d level can hold a maximum of ten electrons so that the ten elements from scandium to zinc inclusively are produced by the development of the 3d penultimate level. These elements form the first series of d-block elements. Most of these have the structure $4s^2$ for their outer shell though there are some exceptions, as shown in Table 9.1. The

Sc	Ti	V	Cr*	Mn	Fe	Co	Ni	Cu*	Zn
$3d^1\,4s^2$	$3d^2\,4s^2$	$3d^3\,4s^2$	$3d^5\,4s^1$	$3d^5\,4s^2$	$3d^6\,4s^2$	$3d^7\,4s^2$	$3d^8\,4s^2$	$3d^{10}\,4s^1$	$3d^{10}\,4s^2$

Table 9.1

exceptions are chromium and copper, which have only one electron in the 4s level. They have the structures shown, in the ground state, because these represent particularly stable configurations. Chromium has a half-filled 3d level and a half-filled 4s level, while copper has a completed 3d level and a half-filled 4s level.

The term *transition element* is sometimes used for the d-block elements, though, more commonly, transition element refers to those elements with a *partially* completed penultimate d level. Strictly speaking, according to this definition only the elements scandium to nickel would be classed as transition elements. However, because copper has an oxidation state of $+2$ in many of its compounds (which gives only nine electrons in its 3d level) these have many properties in common with transition element compounds, and so the term is usually extended to cover copper. Some of the periodic properties of the d-block elements are given in Table 9.2.

I T_d	II T_d	III T_d	IV T_d	V T_d	VI T_d	VII T_d	VIII T_d	IX T_d	X T_d
Sc	Ti	V	Cr	Mn	Fe	Co	Ni	Cu	Zn
$3d^1\,4s^2$	$3d^2\,4s^2$	$3d^3\,4s^2$	$3d^5\,4s^1$	$3d^5\,4s^2$	$3d^6\,4s^2$	$3d^7\,4s^2$	$3d^8\,4s^2$	$3d^{10}\,4s^1$	$3d^{10}\,4s$
0·144 | 632	0·136 | 661	— | 653	— | 653	— | 715	— | 761	— | 757	— | 736	0·138 | 745	0·131 | 9
0·162 | 1·3	0·147 | 1·5	0·134 | 1·6	0·127 | 1·6	0·126 | 1·5	0·126 | 1·8	0·125 | 1·8	0·124 | 1·8	0·128 | 1·9	0·138 | 1
0·081(+3)	0·090(+2) 0·068(+6)	0·074(+3) 0·059(+5)	0·069(+3) 0·052(+6)	0·080(+2) 0·046(+7)	0·076(+2) 0·064(+3)	0·078(+2) 0·063(+3)	0·078(+2) 0·062(+3)	0·096(+1) 0·069(+2)	0·074(+

Y	Zr	Nb	Mo	Tc	Ru	Rh	Pd	Ag	Cd
$4d^1\,5s^2$	$4d^2\,5s^2$	$4d^4\,5s^1$	$4d^5\,5s^1$	$4d^5\,5s^2$	$4d^7\,5s^1$	$4d^8\,5s^1$	$4d^{10}\,5s^0$	$4d^{10}\,5s^1$	$4d^{10}\,5$
0·162 | 636	0·148 | 664	— | 653	— | 694	— | 699	— | 724	— | 745	— | 803	0·153 | 732	0·148 | 8
0·180 | 1·3	0·160 | 1·4	0·146 | 1·6	0·139 | 1·8	0·136 | 1·9	0·134 | 2·2	0·134 | 2·2	0·137 | 2·2	0·144 | 1·9	0·154 | 1
0·093(+3)	0·080(+4)	0·070(+5)	0·068(+4) 0·062(+6)	—	0·069(+3) 0·065(+4)	0·086(+2)	0·050(+2)	0·126(+1)	0·097(+

La	Hf	Ta	W	Re	Os	Ir	Pt	Au	Hg
$5d^1\,6s^2$	$4f^{14}\,5d^2$ $6s^2$	$4f^{14}\,5d^3$ $6s^2$	$4f^{14}\,5d^4$ $6s^2$	$4f^{14}\,5d^5$ $6s^2$	$4f^{14}\,5d^6$ $6s^2$	$4d^{14}\,5d^7$ $6s^2$	$4f^{14}\,5d^{10}$ $6s^0$	$4f^{14}\,5d^{10}$ $6s^1$	$4f^{14}\,5d^x$ $6s^2$
0·169 | 540	— | 531	— | 577	— | 770	— | 761	— | 841	— | 887	— | 866	0·150 | 841	0·149 | 1
0·187 | 1·1	0·158 | 1·3	0·146 | 1·5	0·139 | 1·7	0·137 | 1·9	0·135 | 2·2	0·136 | 2·2	0·138 | 2·2	0·144 | 2·4	0·157 | 1
0·115(+3)	0·081(+4)	0·073(+5)	0·064(+4) 0·068(+6)	—	0·067(+4)	0·066(+4)	0·052(+2)	0·137(+2)	0·110(+

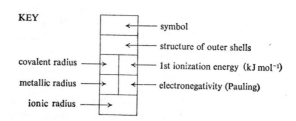

KEY

covalent radius ⟶
metallic radius ⟶
ionic radius ⟶

⟵ symbol
⟵ structure of outer shells
⟵ 1st ionization energy (kJ mol⁻¹)
⟵ electronegativity (Pauling)

Table 9.2

Structural characteristics

(i) Most of the d-block elements have the electronic structure ns^2 for their outer shell. It is usually the structure of the penultimate d sub-level which varies from element to element.

(ii) Palladium and platinum (Group VIIIT_d) have a completed penultimate d level and no electrons in the outer shell. Copper, silver and gold (Group IXT$_d$) have a completed penultimate d sub-level and one electron in their outer s levels. Zinc, cadmium and mercury (Group XT$_d$) have completed d levels and a complete outer s sub-level.

(iii) There is not a large energy difference between the penultimate d level and the outer s level.

(iv) The penultimate d electrons have a relatively poor screening effect so that the nuclear charge has a relatively large effect at the periphery of the atom.

(v) The first ionization energies vary from 531 to 1008 kJ mol^{-1}, and each atom has a relatively large number of empty sub-levels in its valence shell; i.e. the conditions for metallic bonding in the elemental state are fulfilled.

(vi) The first ionization energies increase relatively slowly across each d-block series. In addition they vary rather irregularly down each group. They tend to decrease down the group for the first three groups and then to increase slowly down the group for the remaining groups.

(vii) The electronegativities have intermediate values, tending to increase across a series to a maximum in the region of the coinage metals and then decrease from the coinage metals to the zinc group.

(viii) The covalent, ionic and metallic radii are not very large and decrease slowly across each series.

General chemistry

Since, in general, electrons are added to the penultimate d level rather than the outer shell in going across a d-block series, there is a more gradual change in properties across the series than in a short period. The d-block elements as a whole will have certain properties in common, which arise from the similar structures of the outer shells and the possession of an incomplete penultimate d sub-level. Elements in the same group, however, have, in general, similar structures for the outer *two* shells and so will show a similarity of properties. Those properties which characterize d-block elements as a whole will first of all be discussed, followed by the discussion of Groups IXT$_d$ and XT$_d$ as examples of vertical relationships.

General properties of the d-block elements

Physical properties

The conditions for metallic bonding are fulfilled (see (v) above) so that physically the d-block elements are typical metals. They are malleable, ductile, relatively good conductors of heat and electricity, of high tensile strength, lustrous, and so on. Their boiling points are plotted against atomic number in Fig. 9.1. The boiling points are all above 2000°C

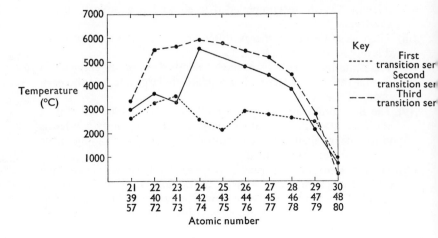

Fig. 9.1

except for the last element in each series, viz. zinc, cadmium and mercury. These latter elements have the structures:

$$Zn \ (Ar) \ 3d^{10} \ 4s^2$$
$$Cd \ (Kr) \ 4d^{10} \ 5s^2$$
$$Hg \ (Xe) \ 4f^{14} \ 5d^{10} \ 6s^2$$

in which the penultimate d level is complete and the outer s sub-level is complete, i.e. all electrons are paired. They will therefore show less tendency to use their outer electrons for bonding; that is, metallic bonds between their atoms will be weaker than for the other elements in each series, so there is a sharp drop in boiling point. The effect will increase down the group, since the outer s electrons show an increasing tendency to remain paired (inert pair effect) as nuclear charge increases. The boiling points decrease down the group.

Variable oxidation state (variable valency)

Because there is not a large energy difference between the outer s sub-level and the penultimate d sub-level most of the d-block elements can use both s and d electrons for bonding. This gives rise to variable oxidation states. The oxidation states for elements in the first series are shown in Table 9.3. The maximum oxidation state increases to a maximum of 7 for manganese and then falls off until it reaches 2 for zinc. This pattern is related to the electronic structures of the elements. The lowest energy electrons are the s electrons, so these may be unpaired and used for

Element	Sc	Ti	V	Cr	Mn	Fe	Co	Ni	Cu	Zn
Oxidation state	(2) 3	2 3 4	2 3 4 5	1 2 3 4 5 6	2 3 4 5 6 7	2 3 4 5 6	2 3 4 5	2 3 4 (5)	1 2 3	2
Total no. of electrons in outer s and penultimate d level	3	4	5	6	7	8	9	10	11	12

Table 9.3

bonding. All the first series elements, therefore, have oxidation states of 2. The maximum oxidation state is then given by:

number of s electrons + theoretical number of unpaired d electrons
i.e.   2 + theoretical number of unpaired d electrons

This is illustrated in Table 9.4. Copper and chromium actually have only

Element	Electronic structure		Maximum oxidation state
	3d orbitals	2s	
Sc	↑	↑↓	2+1 = 3
Ti	↑ ↑	↑↓	2+2 = 4
V	↑ ↑ ↑	↑↓	2+3 = 5
Cr	↑ ↑ ↑ ↑	↑↓ *	2+4 = 6
Mn	↑ ↑ ↑ ↑ ↑	↑↓	2+5 = 7
Fe	↑↓ ↑ ↑ ↑ ↑	↑↓	2+4 = 6
Co	↑↓ ↑↓ ↑ ↑ ↑	↑↓	2+3 = 5
Ni	↑↓ ↑↓ ↑↓ ↑ ↑	↑↓	2+2 = 4
Cu	↑↓ ↑↓ ↑↓ ↑↓ ↑	↑↓ *	2+1 = 3
Zn	↑↓ ↑↓ ↑↓ ↑↓ ↑↓	↑↓	2+0 = 2

* For these elements the ground state structures are more accurately represented as

Cr | ↑ | ↑ | ↑ | ↑ | ↑ | | ↑ |

Cu | ↑↓ | ↑↓ | ↑↓ | ↑↓ | ↑↓ | | ↑ |

Table 9.4

one electrón in the outer s level in the ground state. This accounts for their minimum oxidation states of 1.

Non-stoichiometry and catalytic activity

The variable oxidation states of the transition elements lead to the formation of non-stoichiometric compounds in which some of the cations are in a higher oxidation state, so that anion sites are left unoccupied to preserve electrical neutrality. Iron (II) and nickel (II) oxides exhibit non-stoichiometry of this type (see p. 87). Analysis of iron (II) oxide shows that its composition varies from $Fe_{0.94}O$ to $Fe_{0.84}O$.

Again, the variable oxidation states of the transition elements probably account for their catalytic activity and that of their compounds. It is possible for a transition element compound to form an unstable intermediate by the element changing its oxidation number. If the intermediate compound so formed enables the reaction to follow a path of lower energy of activation, then the reaction velocity will be increased. The transition metals also act as surface catalysts. The atoms at the surface of the metal will possess empty or partly-filled d orbitals which may be used for bond formation. This provides a mechanism for the absorption of gaseous reactants on the surface of the metal, again reducing the activation energy and increasing the velocity of reaction.

Complex formation, formation of coloured compounds and paramagnetism

Complex formation

In general the d-block elements show a marked tendency to form complex ions. The ionic radii of the d-block metals are relatively small and the penultimate d electrons have a relatively poor screening effect; thus there is a tendency to attract electron-donating groups and ions, and so form a complex. The outer groups (ligands) may be bonded to the central atom by means of valence shell orbitals only, or by means of valence shell orbitals and penultimate d orbitals. The bonding in the tetramminezinc ion illustrates the use of valence orbitals only.

Zinc has the structure $Zn [Ar] 3d^{10} 4s^2$, so the zinc ion may be represented as follows:

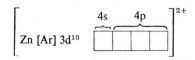

Hybridization of the empty 4s orbital and the three empty 4p orbitals then gives four equivalent sp^3 hybrid orbitals, tetrahedrally orientated.

The donation of the lone pairs from four ammonia molecules into these empty hybrid orbitals then gives the tetramminezinc complex,

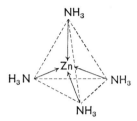

The use of penultimate d orbitals in bonding is illustrated by the formation of the hexacyanoferrate (III) (ferricyanide) ion. Iron has the structure Fe [Ar] $3d^6 4s^2$, so the structure of the Fe^{3+} ion may be represented as:

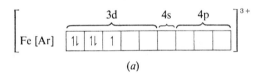

(a)

Hybridization of the two empty 3d orbitals, the single empty 4s orbital and the three empty 4p orbitals may then occur to give six equivalent octahedrally orientated d^2sp^3 hybrid orbitals. Electrons from six cyanide ions may then be donated into the empty hybrid orbitals to give the octahedral hexacyanoferrate (III) ion:

$$\begin{bmatrix} & & CN & & \\ NC & & | & & CN \\ & & Fe & & \\ NC & & | & & CN \\ & & CN & & \end{bmatrix}^{3-}$$

The geometry of complex ions can often be predicted using the electron pair repulsion theory (p. 172). The geometrical shapes associated with various coordination numbers are summarized in Table 9.5.

Coordination number	Geometry
2	linear
3	trigonal
4	tetrahedral (or square planar)
5	trigonal bipyramid
6	octahedral
7	pentagonal bipyramid

Table 9.5

The configuration of the iron (III) ion given above needs some explanation. According to the principle of maximum multiplicity the expected configuration would be

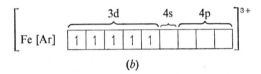

(b)

since the five d orbitals would be expected to be half-filled by single electrons before electron pairing occurred. However, magnetic measurements indicate that in the hexacyanoferrate (III) complex the iron atom has only one unpaired electron, as indicated in configuration (a). That the central iron atom assumes configuration (a) rather than (b) is explained by the *ligand field theory*. This theory is also required for the understanding of colour and paramagnetism in the transition element complexes.

The ligand field theory

The shapes and orientations of the d orbitals have already been discussed in Chapter 1. They are shown again in Fig. 9.2. In Fig. 9.2b the two-lobed d_{z^2} orbital is shown shaded and the four-lobed $d_{x^2-y^2}$ orbital is shown unshaded. The dashed lines (diagonals of the three planes) show the directions of the lobes of the four-lobed d_{xy}, d_{xz} and d_{yz} orbitals.

Consider the formation of the hexacyanoferrate (III) ion. The six cyanide ions eventually assume an octahedral configuration about the central iron (III) ion. The cyanide ions may be considered to approach the iron (III) ion along the x, y and z axes. This means that two of the cyanide ions approach in opposite directions along the axis of the d_{z^2} orbital and four of them approach at right angles along the two axes of the $d_{x^2-y^2}$ orbital, as shown in Fig. 9.3. The cyanide ions possess a nega-

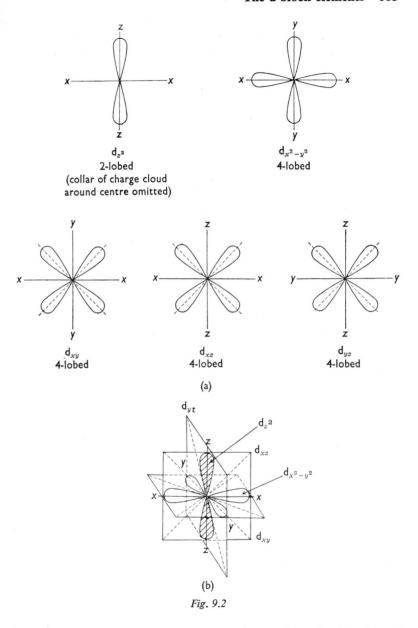

d_{z^2}
2-lobed
(collar of charge cloud
around centre omitted)

d_{x^2-y^2}
4-lobed

d_{xy}
4-lobed

d_{xz}
4-lobed

d_{yz}
4-lobed

(a)

(b)

Fig. 9.2

tive electric field, which will be at a maximum along the direction of approach, i.e. along the x, y and z axes. If the field is strong enough any electrons present in the d orbitals will be repelled and tend to occupy the d_{xy}, d_{yz} and d_{xz} orbitals rather than the d_{z^2}, $d_{x^2-y^2}$ orbitals, since the former group of orbitals does not lie along the directions of approach of the ligands. In other words the field associated with the approaching ligands increases the energy of the d_{z^2}, $d_{x^2-y^2}$ orbitals relative to the

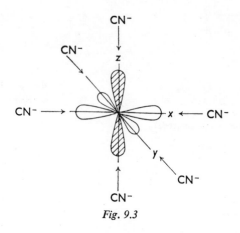

Fig. 9.3

d_{xy}, d_{yz} and d_{xz} orbitals. The five d orbitals, which are of equal energy in the absence of the octahedral ligand field, split into two groups of different energy, the d_γ group (made up of the d_{z^2} and $d_{x^2-y^2}$ orbitals) of relatively high energy, and the d_ϵ group (made up of the d_{xy}, d_{yz} and d_{xz} orbitals) of relatively low energy. This is illustrated in Fig. 9.4.

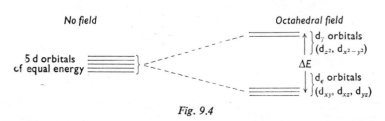

Fig. 9.4

When electrons redistribute themselves among the d orbitals two factors will affect the outcome:

(*a*) The tendency for electrons to get as far away from each other as possible because of their like charge. This results in the half-filling of the d orbitals by single electrons before electron pairing occurs (*principle of maximum multiplicity*).

(*b*) The tendency for orbitals of lower energy to be completely filled before higher energy orbitals are occupied.

The greater the value of ΔE the more important the second factor will become. The cyanide ion produces a strong ligand field, so ΔE is large.

Therefore, in the hexacyanoferrate (III) ion the low-energy d_y orbitals are occupied in preference to the d_ϵ orbitals, thus:

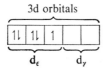

The d_y orbitals, being vacant, are then available for bonding. If the octahedral field is weak then ΔE is small; the first factor therefore is the important one, so that the principle of maximum multiplicity governs the electron distribution. The fluoride ion produces only a weak ligand field, and thus in the hexafluoroferrate (III) ion $[FeF_6]^{3-}$ the iron adopts the following configuration:

This is known as a 'spin-free' configuration, in contrast to that, in the hexacyanoferrate (III) ion, which is referred to as 'spin-paired'. The order of increasing field strength for some common ligands is as follows:

halogens < water < ammonia and amines < cyanide

This corresponds to the order of decrease in electronegativity or increase in ease of polarization.

The ligand field approach can also be applied to the case of tetrahedral coordination. Fig. 9.5 shows the relationship between tetrahedral

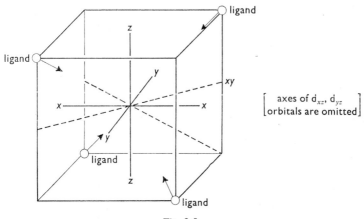

Fig. 9.5

7+

ligand approach and the d-orbital axes. The ligands do not approach the nucleus along the axes of either the d_ϵ or the d_γ orbitals. However, it can be shown that the line of approach makes an angle of 35° 16′ with the direction of a d_ϵ orbital and an angle of 54° 44′ with the direction of a d_γ orbital. The approaching ligands would therefore have a greater repulsive effect on electrons in d_ϵ orbitals than on electrons in d_γ orbitals. For a tetrahedral field, therefore, the d_ϵ orbitals are the ones of higher energy, as shown in Fig. 9.6. Since the ligands do not approach

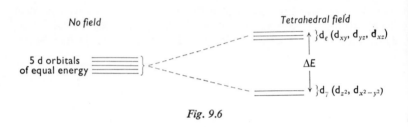

Fig. 9.6

along the axes of the d_ϵ orbitals, and since there are four rather than six ligands, the value of ΔE is less for corresponding ligands than that for an octahedral ligand field.

Paramagnetism

If a substance, when suspended in an inhomogeneous magnetic field, moves from the stronger to the weaker part of the field, it is said to be *diamagnetic*. On the other hand if it moves from the weaker to the stronger part of the field it is said to be *paramagnetic*. Diamagnetic substances have only paired electrons while paramagnetic substances have unpaired electrons. The paramagnetic susceptibility of a substance, from which the paramagnetic moment can be calculated, is determined using a Gouy balance in which the specimen is weighed in the presence and then in the absence of a magnetic field. The value of the magnetic moment is related to the number of unpaired electrons. Many transition element compounds are paramagnetic because the d orbitals often contain unpaired electrons. Measurement of the magnetic moments of potassium hexacyanoferrate (III) and potassium hexafluoroferrate (III) indicates the presence of one and five unpaired electrons respectively, which confirms the structures of the ions predicted by the ligand field theory.

Colour

White light can be split up, e.g. by means of a prism, into a number of different colours, the visible spectrum. Each colour is characterized by a definite wavelength or frequency. The energy associated with a particular colour is given by the equation:

$$E = h\nu$$

where E = the energy of the radiation
h = Planck's constant
ν = the frequency.

But

$$c = \nu\lambda$$

where c = velocity of light
λ = wavelength.

Therefore

$$E = \frac{hc}{\lambda}$$

The wavelength range for the visible spectrum, together with the associated wavenumber (number of vibrations per cm) and energy ranges, is given in Fig. 9.7.

Colour	Wavelength (nm)	Wavenumber, $1/\lambda$ (cm^{-1})	Energy (kJ mol^{-1})
Violet	400	25 000	299
Blue			
	500	20 000	239
Green			
Yellow	600	16 667	195
Orange			
Red	700	14 300	171

Fig. 9.7

When white light passes through a coloured solution (or transparent solid) certain frequencies are absorbed and the solution has the colour of the transmitted light. Energy is absorbed in promoting electrons from lower to higher energy levels. The smallest ionization energy (377 kJ mol^{-1} for caesium) is greater than the energy available from violet light (299 kJ mol^{-1}) so that absorption of visible radiation must be due to intra-atomic electron transitions.

Table 9.6 shows the colours of solutions of the hydrated ions of metals of the first d-block series. Inspection of the table shows that those ions which have closed shell configurations are colourless since electron transition would require the absorption of a very large amount of energy

Electronic structure of ion	Number of unpaired d electrons	Ion (hydrated)	Colour
Sc [Ar] $3d^0$ $4s^0$	0	(K^+, Ca^{2+}) Sc^{3+}	colourless
Ti [Ar] $3d^1$ $4s^0$	1	Ti^{3+}	pink-violet
V [Ar] $3d^2$ $4s^0$	2	V^{3+}	green
Cr [Ar] $3d^3$ $4s^0$	3	Cr^{3+}	violet
Cr [Ar] $3d^4$ $4s^0$	4	Cr^{2+}	blue
Mn [Ar] $3d^5$ $4s^0$	5	Mn^{2+}	very pale pink
Fe [Ar] $3d^5$ $4s^0$	5	Fe^{3+}	very pale violet
Fe [Ar] $3d^6$ $4s^0$	4	Fe^{2+}	green
Co [Ar] $3d^7$ $4s^0$	3	Co^{2+}	pink
Ni [Ar] $3d^8$ $4s^0$	2	Ni^{2+}	green
Cu [Ar] $3d^9$ $4s^0$	1	Cu^{2+}	blue
Zn [Ar] $3d^{10}$ $4s^0$	0	(Cu^+) Zn^{2+} (Ga^{3+})	colourless

Table 9.6

and could not be brought about by radiation in the visible range. Consider the $[Ti(H_2O)_6]^{3+}$ ion as an example of a coloured ion. The six water molecules surround the central titanium (III) ion and give rise to an octahedral ligand field. This causes the d orbitals to split into d_γ orbitals of relatively high energy and d_ϵ orbitals of lower energy. The energy difference between the d_γ and d_ϵ orbitals corresponds to a wavenumber of $20\,000$ cm^{-1}, which lies within the visible spectrum. If the solution is illuminated with white light, absorption of radiation of this wavenumber occurs to effect an electron transition from the d_ϵ to the d_γ orbitals. The solution, therefore, has a characteristic pinkish-violet colour.

The colours of $[Mn(H_2O)_6]^{2+}$ and $[Fe(H_2O)_6]^{3+}$ are very weak. This is because both of these ions have the structure $d_\epsilon^3\,d_\gamma^2$ for their d levels. The orbitals are exactly half-filled, which is a relatively stable structure. Electron transition would involve the excitation of an unpaired electron from a d_ϵ orbital to a d_γ orbital to give two paired electrons in one of the d_γ orbitals, which is a less stable structure. The intensity of the absorption bands is therefore very small.

Any factor which changes the value of ΔE or alters the number or intensity of the absorption bands will alter the colour of the compound. Some of the important factors are:

(a) A change in the number of d electrons owing to a change in the central atom; e.g. the $[Co(H_2O)_6]^{2+}$ ion (d^7) is pink, while the $[Ni(H_2O)_6]^{2+}$ ion (d^6) is green.

(b) A change in the number of d electrons owing to a change in oxidation state of the central atom; e.g. the $[Cr(H_2O)_6]^{2+}$ ion (d^4) is blue, while the $[Cr(H_2O)_6]^{3+}$ ion (d^3) is green. ΔE is also increased because of the increase in ionic charge.

(c) A change in the value of ΔE for different ligands, since the strength of the ligand field depends upon this; e.g. $[Cu(H_2O)_4]^{2+}$ is pale blue, but $[Cu(NH_3)_4]^{2+}$ is violet.

(d) A change in ΔE owing to a change in orientation of the ligands; e.g. a change from octahedral to tetrahedral ligand field. For example, octahedral $[Mn(H_2O)_6]^{2+}$ is pink, while tetrahedral $[MnCl_4]^{2-}$ is green.

(e) A change in ΔE owing to isomerism; e.g. in the octahedral complex $[Co(NH_3)_4Cl_2]^+$ the isomer with the chlorine atoms adjacent (*cis*-form) is violet, while that with the chlorine atoms opposite (*trans*-form) is green.

Not all colour in inorganic compounds arises from transitions between the d orbitals. Silver iodide, for example, is yellow and yet the silver atom has a closed 4d shell. If silver iodide were purely ionic, Ag^+I^-, then one would expect it to be colourless. However, the structure of silver iodide may be represented as ionic, Ag^+I^-, or covalent, $Ag:I$. If the energy difference between these two configurations corresponds with an energy quantum in the visible range, absorption will occur and the substance will be coloured. The yellow colour of silver iodide is thought to arise from this type of charge transfer phenomenon. Charge transfer is probably responsible for the deep colours of many oxides and sulphides and for the intense colours of potassium chromate and potassium permanganate. When potassium hexacyanoferrate (III) and iron (II) sulphate solutions are mixed, an intense blue precipitate of Prussian blue, $KFe[Fe(CN)_6]$, is obtained. Its colour is probably due to charge transfer between the two forms

$$KFe^{II}[Fe^{III}(CN)_6] \quad \text{and} \quad KFe^{III}[Fe^{II}(CN)_6]$$

The first transition series—scandium to nickel

Scandium (Sc [Ar] $3d^1 4s^2$)

Scandium has one 3d electron and two 4s electrons. It uses all three of these for bonding, so it always has an oxidation state of $+3$. The ionic radius of the Sc^{3+} ion is relatively large (0.081 nm); the nuclear charge is relatively low, and it has only completed shells. Therefore, scandium compounds are colourless and the Sc^{3+} ion shows little tendency to form complexes. Scandium resembles aluminium in many ways. Its hydroxide, however, is more basic than aluminium hydroxide, though less basic than calcium hydroxide. It forms a hydride of variable composition. This apparently contains hydride ions since it reacts with water to give hydrogen. This again illustrates its relatively low electronegativity.

Titanium (Ti [Ar] $3d^2 4s^2$)

Titanium shows oxidation states of $+3$ and $+4$, the latter being the more stable. The titanium (IV) ion has no d electrons, so its compounds are colourless. Titanium (III), however, has a single unpaired d electron and its compounds are pinkish violet. The Ti^{4+} ionic radius is relatively small, so simple compounds of titanium (IV) tend to be covalent; e.g. $TiCl_4$, which is a colourless covalent liquid. Titanium (IV) ions readily form complexes with electron-donating molecules such as pyridine or di-ethyl ether. Titanium forms an interstitial hydride and nitride.

Vanadium (V [Ar] $3d^3 4s^2$)

This element has a large number of oxidation states, as shown in its oxides VO, V_2O_3, VO_2, V_2O_5. Vanadium (II) compounds are ionic and resemble iron (II) compounds. Vanadium (III) compounds resemble iron (III) compounds. Vanadium (V) oxide, V_2O_5, has no unpaired electrons and yet is orange in colour; this is probably a result of charge transfer absorption since the material is non-stoichiometric. Vanadium (V) oxide gives rise to vanadates which are analogous to phosphates; e.g. Na_3VO_4 is the analogue of Na_3PO_4. Like the phosphate ion the vanadate ion tends to polymerize. In strongly alkaline solution there are probably simple vanadate ions present but as the acidity increases (pH decreases) polymerization occurs, until in strongly acidic solution the basic character of V_2O_5 comes into effect and vanadyl ions, VO^{3+}, are formed:

$$\underbrace{VO_4^{3-} \xrightarrow{\text{pH11}} V_2O_7^{4-} \xrightarrow{\text{pH9}} H_2V_4O_{13}^{4-}}_{\text{colourless}}$$

$$\downarrow \text{pH7}$$

$$\underbrace{H_4V_5O_{16}^{3-} \xrightarrow{\text{pH2}} V_2O_5(H_2O)_n}_{\text{orange-brown}} \xrightarrow{\text{pH} < 1} \underbrace{VO^{3+}}_{\text{yellow}}$$

Chromium (Cr [Ar] $3d^5 4s^1$)

The usual oxidation states of chromium are $+2$, $+3$ and $+6$. Chromium (II) compounds are usually ionic but are strong reducing agents, readily reverting to the $+3$ state. The $+3$ oxidation state is the most stable, though chromium (III) compounds are readily oxidized to the $+6$ state in alkaline solution. Chromates and dichromates, in which the chromium is in the $+6$ state, are strong oxidizing agents. Chromium (II) hydroxide, $Cr(OH)_2$, is basic, chromium (III) oxide, Cr_2O_3, is amphoteric and chromium (VI) oxide, CrO_3, is strongly acidic. Since chromium (like the other members of this group) has six outer electrons it has some resemblance to Group VIM elements which also have six outer electrons. For example sulphur (VI) oxide, SO_3, is highly acidic, like chromium (VI) oxide, CrO_3. The ions derived from these oxides have similar structures:

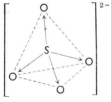

tetrahedral sulphate ion

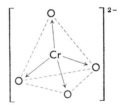

tetrahedral chromate ion

Unlike sulphate ions, however, chromate ions show a strong tendency to polymerize in acid solution. For example,

$$2CrO_4^{2-} \underset{}{\overset{2H^+}{\rightleftharpoons}} Cr_2O_7^{2-} + H_2O$$
$$\text{dichromate ion}$$

Although chromium does not form any chromium (VI) halides, it forms a chromium (VI) oxychloride (chromyl chloride) CrO_2Cl_2. This is a dark red liquid made by heating a chloride with potassium dichromate and concentrated sulphuric acid. Chromium forms a trichloride and a tribromide, which have layer structures in the anhydrous state. The hydrated trichloride exhibits hydrate isomerism, there being three different forms:

$$[Cr(H_2O)_6]Cl_3 \quad \text{violet}$$
$$[Cr(H_2O)_5Cl]Cl_2.H_2O \quad \text{pale green}$$
$$[Cr(H_2O)_4Cl_2]Cl.2H_2O \quad \text{dark green}$$

Only the chloride ion which is not involved in the complex can be precipitated with silver nitrate. The Cr^{3+} ion forms many complexes, most of which are octahedral, e.g. $[Cr(NH_3)_6]^{3+}$ and $[CrCl_6]^{3-}$. Like many tripositive ions the Cr^{3+} ion forms an alum, $KCr(SO_4)_2.12H_2O$.

Manganese (Mn [Ar] $3d^5 4s^2$)

The metal is very similar to metallic iron. It reacts with dilute acids to give hydrogen. It reacts with the halogens to give di-halides (except for fluorine which gives the trifluoride, MnF_3) and with sulphur and oxygen to give manganese (II) sulphide and manganese (III) oxide respectively. The maximum oxidation state of manganese is $+7$, as in the purple permanganate ion MnO_4^-, but its most common oxidation state is $+2$, as in manganese (II) salts, e.g. $MnSO_4$. Hydrated manganese (II) ions are pink, and have enhanced stability because the 3d sub-level is exactly half-filled. The manganese (II) ion readily forms complexes, e.g. $[Mn(CN)_6]^{4-}$. The cyanide complex can be relatively easily oxidized and reduced to give manganese in the less common oxidation states of $+3$ and $+1$:

$$[Mn(CN)_6]^{5-} \xleftarrow[\text{Zn}]{\text{reduction}} [Mn(CN)_6]^{4-} \xrightarrow[\text{air}]{\text{oxidation}} [Mn(CN)_6]^{3-}$$

$$\text{Mn}(+1) \qquad\qquad\quad \text{Mn}(+2) \qquad\qquad\quad \text{Mn}(+3)$$

In the green manganate ion, MnO_4^{2-}, the manganese has an oxidation state of $+6$. In acid solution this ion disproportionates to give manganese dioxide and permanganate ions:

$$3MnO_4^{2-} + 4H^+ \longrightarrow MnO_2 + 2MnO_4^- + 2H_2O$$

$$\text{Mn}(+6) \qquad\qquad\quad \text{Mn}(+4) \quad \text{Mn}(+7)$$

Because manganese has seven outer electrons it has a slight resemblance to the halogens in Group VIIM. For instance, manganese (VII) oxide, Mn_2O_7, and chlorine (VII) oxide, Cl_2O_7, are both strongly acidic and give rise to permanganate ions, MnO_4^-, and perchlorate ions, ClO_4^-, respectively. These ions have similar tetrahedral structures, and compounds containing them, e.g. $KMnO_4$ and $KClO_4$, are isomorphous, though permanganates are purple whereas perchlorates are colourless. Manganese resembles its horizontal neighbours in the transition series more strongly than the halogens.

Iron (Fe [Ar] $3d^6$ $4s^2$)

Iron has a maximum oxidation number of $+6$ in the ferrate ion, FeO_4^{2-}. This is a powerful oxidizing agent, however; the most stable oxidation states of iron being iron (II) $(+2)$ and iron (III) $(+3)$. The stability of the iron (II) and iron (III) states is to be expected, since in the former only s electrons are used and in the latter the 3d orbitals are exactly half-filled. Iron has zero oxidation state in the carbonyls $Fe(CO)_5$ and $Fe(CO)_{12}$.

Iron (II) ions form complexes, such as the yellow hexacyanoferrate (II) ion and the brown complex formed in the brown-ring test for nitrates, which is probably $[Fe(NO)]^{2+}$. Iron (II) compounds are more ionic than iron (III) compounds, since the iron (II) ion has a larger radius and a smaller charge. Iron (II) hydroxide, therefore, is a stronger base than iron (III) hydroxide, and iron (III) fluoride, which might be expected to be ionic, does not give positive tests for iron (III) or fluoride ions in aqueous solution.

Iron reacts directly with chlorine, when heated, to give anhydrous iron (III) chloride, which is dimeric in the gaseous state:

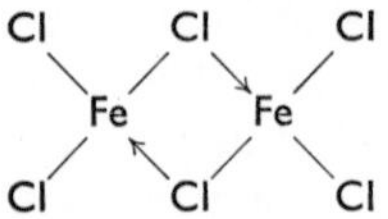

Iron (III) ions form complex cyanide ions, $[Fe(CN)_6]^{3-}$, and in the ammonium thiocyanate test for a ferric salt the deep red complex $[Fe(SCN)]^{2+}$ is formed.

Cobalt (Co [Ar] $3d^7\ 4s^2$)

Like iron, cobalt commonly has oxidation states of $+2$ and $+3$. Cobalt (II) ions form both tetrahedral complexes such as $[CoCl_4]^{2-}$ and octahedral complexes such as $[Co(H_2O)_6]^{2+}$. The cobalt (III) ion is a strong oxidizing agent and will oxidize water to oxygen. However, it is stabilized by complexing and in fact forms more complexes than any other cation. It forms stable cation, anion and neutral complexes, such as

$$[Co(NH_3)_6]^{3+}, \quad [Co(NO_2)_6]^{3-} \quad \text{and} \quad [Co(NH_3)_3(NO_2)_3]^0$$

which are all yellow. Its complexes exhibit a number of different types of isomerism (see p. 217), such as the following:

(i) *Coordination isomerism*

$$[Co(NH_3)_6]^{3+}[Cr(SCN)_6]^{3-} \quad \text{and} \quad [Co(NH_3)_4(SCN)_2]^+[Cr(NH_3)_2(SCN)_4]^-$$

(ii) *Hydrate isomerism*

$$[Co(NH_3)_3(H_2O)_2Cl]^{2+}2Br^- \quad \text{and} \quad [Co(NH_3)_3(H_2O)ClBr]^+Br^-.H_2O$$

(iii) *Geometrical isomerism*

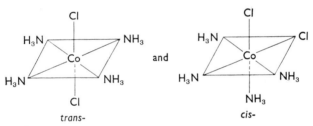

(iv) *Optical isomerism*

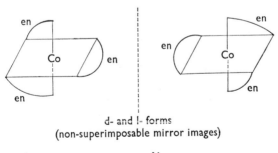

d- and l- forms
(non-superimposable mirror images)

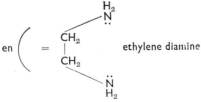

7*

Nickel (Ni [Ar] $3d^8 4s^2$)

Although, like iron and cobalt, nickel may have oxidation states of $+2$ and $+3$, the former is the more stable, and in most of its compounds nickel is in the $+2$ state. Like cobalt the nickel (II) ion readily forms complexes, which may be octahedral, such as $[Ni(H_2O)_6]^{2+}$, tetrahedral, such as $[NiCl_4]^{2-}$, or square planar, such as $[Ni(CN)_4]^{2-}$. The first two are formed when the ligands exert a relatively weak ligand field. Square planar complexes are formed with groups exerting a strong ligand field. The structure of the nickel (II) ion is:

$$(Ni [Ar] 3d^8 4s^0)^{2+}$$

so that the structure of the 3d sub-level in the absence of a ligand field would be

3d orbitals

⇅	⇅	⇅	↑	↑

If six cyanide ions approach the nickel (II) ion octahedrally, the four which approach along the axes of the $d_{x^2-y^2}$ orbital exert a strong repulsive effect on the electron present in this orbital, repelling it into the d_{z^2} orbital where it pairs up with the electron already present. The pair of electrons in the d_{z^2} orbital then have a strong repulsive effect on the other two approaching cyanide ions, so that these are repelled away. The result is a square planar complex,

3d   4s   4p

⇅	⇅	⇅	⇅	☒		☒		☒	☒

dsp^2 hybridization

where $\times$ is an electron pair donated by the ligand.

Nickel (II) ions are detected by the red precipitate obtained when dimethylglyoxime is added to a slightly ammoniacal solution of a nickel (II) salt. This is a precipitate of nickel bis-dimethylglyoxime,

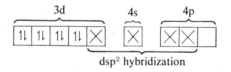

The metals of Groups IXT$_d$ and XT$_d$

GROUP IXT$_d$; THE COINAGE METALS,
COPPER, SILVER AND GOLD

Structural characteristics

(i) All three metals, copper, silver and gold, have the structure $(n-1)d^{10} ns^1$ for their outer two sub-levels.

(ii) The first ionization energies show a slight decrease from copper to silver and then a relatively sharp increase.

(iii) The electronegativities are relatively high. The values are the same for copper and silver but gold shows a relatively sharp increase.

General chemistry

All three elements show the characteristic properties of metals. They are lustrous, malleable, ductile and very good conductors of heat and electricity.

The relatively high electronegativities and first ionization energies give rise to the following properties:

(a) They are displaced from solutions of their salts by more electronegative metals such as zinc.

(b) They are attacked directly only by the very highly electronegative elements such as chlorine. Gold chloride is decomposed readily by heating.

(c) Only copper is attacked directly by oxygen. Silver and gold oxides must be prepared by indirect methods and they are readily decomposed by heating.

(d) They do not displace hydrogen from acids. Copper and silver are attacked by oxidizing acids, gold by aqua regia only.

Since they all have a completed d level and one electron in the outer s level, the coinage metals all exhibit an oxidation state of +1. The d electrons can also be used for bonding, so the Group IXT$_d$ elements exhibit variable oxidation states. The +1 compounds have only completed shells and so are colourless. The higher oxidation states have incomplete d levels and so are coloured:

$$\begin{array}{lll} \text{copper} & \text{(II)} & \text{blue} \\ \text{silver} & \text{(II)} & \text{brown} \\ \text{gold} & \text{(III)} & \text{brown} \end{array}$$

The relatively high nuclear charges and the poor screening effects of the penultimate d electrons have the following consequences,

(1) The coinage metals tend to form covalent rather than ionic compounds; e.g. copper (I) chloride, Cu_2Cl_2, is dimeric in the vapour

state, is only partially ionized in the fused state and does not have an ionic crystal lattice. There is no evidence for the existence of simple copper (II) or gold (II) ions and only one known case of the existence of a silver (II) ion, viz. in silver (II) fluoride, AgF_2.

(2) The coinage metal ions tend to form complexes with electron-donating groups or ions, e.g. $[Cu(NH_3)_4]^{2+}$.

Silver shows somewhat anomalous behaviour compared to copper and gold in that silver (I) compounds are more stable than copper (I) and gold (I) compounds. For example, silver (I) nitrate and sulphate are quite stable in aqueous solution whereas the analogous copper (I) and gold (I) salts break down to give the corresponding copper (II) and gold (III) compounds. For example,

$$Cu_2SO_4 \xrightarrow{H_2O} Cu + CuSO_4$$
$$Cu\,(+1) \qquad\qquad Cu\,(+2)$$

Again, only two silver (II) compounds are known; they are silver (II) fluoride, AgF_2, and silver (II) oxide, AgO.

The coinage metals resemble the alkali metals structurally, both groups having one outer s electron, other levels being closed. They resemble the alkali metals chemically in that they form colourless metal (I) compounds, but otherwise the resemblance is slight.

The copper (I) ion forms tetrahedral cationic and anionic complexes, such as the following:

Copper (II) forms both cationic and anionic complexes, e.g.

$$[Cu(NH_3)_4]^{2+} \quad \text{and} \quad [CuCl_4]^{2-}$$

The tetramminecopper (II) ion is square planar, the electronic arrangement being uncertain.

Copper (II) compounds containing weakly electronegative radicals decompose spontaneously at room temperature to give the corresponding copper (I) compound; for example, if potassium iodide solution is added to copper (II) sulphate solution copper (I) iodide is precipitated:

$$CuSO_4 + 2KI \longrightarrow (CuI_2) + K_2SO_4$$
$$2(CuI_2) \longrightarrow Cu_2I_2 + I_2$$

Copper (II) compounds containing highly electronegative radicals decompose on heating to give the copper (I) compound:

$$2CuCl_2 \xrightarrow{\text{heat}} Cu_2Cl_2 + Cl_2$$

Insoluble copper (I) compounds are stable in contact with water, but other copper (I) compounds disproportionate in the presence of water:

$$2Cu^+ \longrightarrow Cu^{2+} + \downarrow Cu(s)$$

Silver (I) forms cationic and anionic complexes, e.g. $[Ag(NH_3)_2]^+$ and $[Ag(S_2O_3)_2]^{3-}$, which are linear. Gold (I) forms the well-known cyanide complex $[Au(CN)_2]^-$, which is produced in the extraction of the metal; gold (III) forms $[AuCl_4]^{3-}$, which is produced when gold is dissolved in aqua regia.

GROUP XT_d; ZINC, CADMIUM AND MERCURY

Structural characteristics

(1) All the elements in the group have the structure $(n-1)d^{10} ns^2$ for their outer two sub-levels.
(2) Their first ionization energies show a small decrease from zinc to cadmium and then a relatively sharp increase from cadmium to mercury. This is because of the relatively sharp increase in nuclear charge of 32 units, from cadmium to mercury.
(3) The elements have moderately high electronegativities which increase from zinc to mercury.

General chemistry

The zinc group elements, like the alkaline earths, have two outer electrons. The two groups resemble each in that they both form colourless metal (II) compounds. Otherwise there is little chemical similarity between them apart from the resemblances between magnesium and zinc discussed in Chapter 15 (p. 114).

All the elements in the group use both outer electrons for bonding. None of the electrons in the penultimate d shell is used and compounds in which only one s electron is used are unknown. (Even in mercury (I) compounds, where the mercury has an oxidation number of $+1$, both electrons are used for bonding, as will be seen later.)

The zinc group elements have relatively high nuclear charges and relatively small ionic radii, and their penultimate d electrons have a relatively poor screening effect. These factors give rise to the following properties:

(a) They often tend to form covalent compounds.
(b) They readily form complex ions.

(*c*) Their oxides are readily reduced to the metal, zinc and cadmium oxide by heating with carbon, and mercury (II) oxide simply by heating.

The chemistry of zinc is similar to that of cadmium, but mercury shows many anomalous properties. Zinc forms compounds which are mainly ionic such as zinc carbonate, $ZnCO_3$; but many of its simple compounds are covalent, such as anhydrous zinc chloride, $ZnCl_2$. The halides dissolve in water to give hydrated zinc ions, $[Zn(H_2O)_4]^{2+}$. Zinc readily forms complexes such as the tetrahedral tetramminezinc (II) ion, $[Zn(NH_3)_4]^{2+}$. Zinc oxide is amphoteric, and the hydroxide first precipitated when sodium hydroxide is added to a solution of a zinc salt dissolves in excess owing to the formation of complex ions:

$$[Zn(H_2O)_4]^{2+} \xrightarrow{OH^-} [Zn(H_2O)_3(OH)]^+ \xrightarrow{OH^-} [Zn(H_2O)_2(OH)_2]^0$$
white precipitate

$$[Zn(OH)_4]^{2-} \xleftarrow{OH^-} [Zn(H_2O)(OH)_3]^- \xleftarrow{OH^-}$$
'zincate' ions

Cadmium oxide on the other hand is basic, and the precipitate of cadmium hydroxide obtained when caustic soda is added to a solution of a cadmium salt is insoluble in concentrated sodium hydroxide.

Cadmium salts are less hydrated than the corresponding zinc salts. Cadmium iodide solution shows anomalous properties owing to self-complexing, with the formation of such species as CdI^+, CdI_2, CdI_3^- and CdI_4^-. Mercury (II) salts are usually anhydrous and show little tendency to ionize. Cadmium iodide is a soft flaky solid having a layer structure; the arrangement of atoms in one layer is shown in Fig. 9.8a. Since the upper and lower surfaces of each layer are made up of negatively charged iodine atoms (Fig. 9.8b) the forces between adjacent

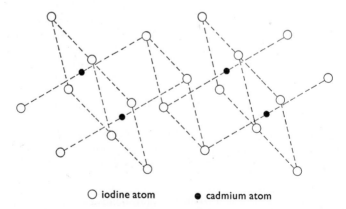

○ iodine atom   ● cadmium atom

Fig. 9.8a   Structure of cadmium iodide layer

layers are relatively weak, which accounts for the flaky character of the solid. The layers are stacked so that the cadmium atoms lie vertically above each other.

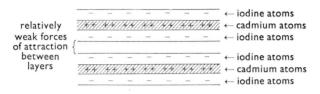

relatively weak forces of attraction between layers { ← iodine atoms ← cadmium atoms ← iodine atoms ← iodine atoms ← cadmium atoms ← iodine atoms

Fig. 9.8b

Mercury has a relatively high nuclear charge so that the inert pair effect tends to be at a maximum. The two s electrons of mercury show a higher degree of inertness than those of zinc and cadmium. This leads to some rather anomalous properties:

(i) Mercury is very volatile for an element in its position in the Periodic Table. Its vapour is monatomic.
(ii) It forms covalent rather than ionic bonds.
(iii) The mercury atom usually forms only two covalent links. Its valence shell seldom holds more than four electrons. Thus, although mercury, like zinc and cadmium, forms complexes with four ligands, two ligands are held more strongly.
(iv) It forms a unique type of ion in the $(+1)$ oxidation state which has the structure

$$(Hg—Hg)^{2+}$$

There is a considerable amount of evidence for the existence of this ion in mercury (I) compounds, the most important being:

(*a*) *Measurement of colligative properties*, e.g. depression of freezing point. This indicates that mercury (I) salts ionize as below:

$$Hg_2X_2 \rightleftharpoons Hg_2^{2+} + 2X^-$$

rather than

$$2HgX \rightleftharpoons 2Hg^+ + 2X^-$$

(*b*) *Measurement of equilibrium constants.* When a mercury (II) salt reacts with mercury the two possibilities are:

$$Hg^{2+} + Hg \rightleftharpoons 2Hg^+ \quad \text{and} \quad Hg^{2+} + Hg \rightleftharpoons Hg_2^{2+}$$

$$K_1 = \frac{[Hg^+]^2}{[Hg^{2+}][Hg]} \qquad K_2 = \frac{[Hg_2^{2+}]}{[Hg^{2+}][Hg]}$$

However, [Hg] is constant, since the concentration of metallic mercury equals its density in $mol\,dm^{-3}$, which is a constant at constant temperature. Therefore,

$$K_1' = \frac{[Hg^+]^2}{[Hg^{2+}]} \quad \text{and} \quad K_2' = \frac{[Hg_2^{2+}]}{[Hg^{2+}]}$$

K_2' turns out to be a constant; hence the mercury (I) ion must be Hg_2^{2+}.

(c) *Measurement of electrical conductivity.* The conductivity of mercury (I) nitrate indicates that it is a ternary electrolyte which suggests that it ionizes, thus:

$$Hg_2NO_3 \rightleftharpoons Hg_2^{2+} + 2NO_3^-$$

(d) *Crystal structure.* X-ray diffraction studies show that mercury (I) chloride crystals contain linear arrangements, Cl–Hg–Hg–Cl, which suggests the presence of the Hg–Hg bond.

If mercury (I) chloride is Hg_2Cl_2 and not HgCl it would be expected that vapour density determination would indicate the appropriate molecular weight. Early determinations appeared to indicate HgCl. However, if the determination of vapour density is carried out under strictly anhydrous conditions a result in accordance with $Hg_2\,Cl_2$ is obtained. The initial result arose from the dissociation of the mercury (I) chloride at high temperatures:

$$Hg_2\,Cl_2 \rightleftharpoons HgCl_2 + Hg$$

That this occurs, was shown by placing a piece of gold foil in contact with the vapour. It was amalgamated by the mercury vapour present.

Mercury (II) compounds are often covalent, e.g. mercury (II) chloride, Cl–Hg–Cl, but some are ionic, e.g. $Hg(NO_3)_2$. Mercury (II) complexes tend to be more stable than those of zinc and cadmium. The best known example of a mercury (II) complex is probably the tetraiodomercurate (II) ion, $[HgI_4]^{2-}$, which is present in Nessler's reagent. Mercury (I) compounds are ionic since they contain the Hg_2^{2+} ion. The mercury (I) ion does not form complexes, probably because of its large size. Most mercury (I) compounds are insoluble and can be prepared by reacting the corresponding mercury (II) compounds with mercury.

References

Books

Chemistry Today, Organisation for Economic Co-operation and Development
Inorganic Chemistry: An Advanced Textbook, T. Moeller (Wiley)
Concise Inorganic Chemistry, J. D. Lee (Van Nostrand)
Structural Principles in Inorganic Compounds, W. E. Addison (Longmans)

Papers

'A Summer Short Course in Co-ordination Chemistry', G. G. Schlessinger (reprint no. 47, from *Chemistry*, June, July, August 1966)

Films

Vanadium—A Transition Element, R. Brasted—a CHEM Study Film, available from Sound-Services Ltd., cat. no. 4172/999

10. Isomerism

Isomerism in organic compounds

Isomerism is the occurrence of two or more compounds which have the same kind and number of atoms, but which differ in the way in which the atoms are arranged. The compounds, called *isomers*, have different properties.

The existence of a very large number of organic compounds with a wide range of different properties depends on isomerism. It is convenient to divide isomerism into two main classes, structural isomerism and stereoisomerism.

(a) STRUCTURAL ISOMERISM

Structural isomers differ in the way in which atoms or groups of atoms are bonded in the molecule. The differences in the structure of isomeric molecules can be shown adequately using planar formulae, if it is understood that the molecules are actually three-dimensional. (The four carbon valency bonds in aliphatic hydrocarbons, for example, are at the tetrahedral angle of 109° 28′ to each other.) Some examples of structural isomers are shown below.

1. Compounds of molecular formula C_4H_{10}

Structural isomers:

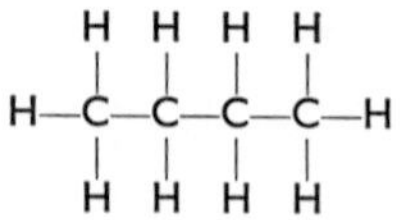

butane 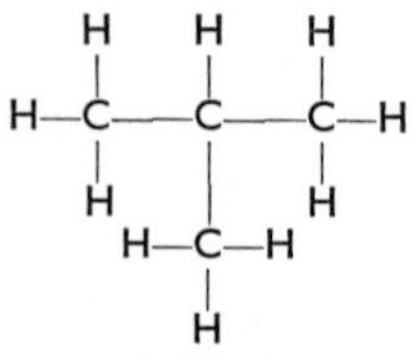   isobutane

2. Compounds of molecular formula C_3H_8O

Structural isomers:

H—C—C—C—O—H
propyl alcohol

H—C—C—C—H
isopropyl alcohol

H—C—C—O—C—H
methyl ethyl ether

3. Compounds of molecular formula $C_8H_8O_2$

Structural isomers:

CH_3 COOH	CH_3 COOH	CH_3 COOH	CH_2COOH
ortho-	*meta-*	*para-*	phenylacetic acid
	toluic acids		

(b) STEREOISOMERISM

Stereoisomers have the same atoms or groups bonded by similar bonds, but differ because corresponding atoms or groups occupy different positions in space. There are two types of isomerism which depend on this spatial factor, geometrical isomerism and optical isomerism.

1. Geometrical isomerism

This depends on the restricted rotation about a double bond. Ethylene has been shown (p. 44) to have the structure illustrated (Fig. 10.1), where the atoms all lie in the same plane. The π-bond is formed by maximum sideways overlap of the p orbitals of the carbon atoms, so that rotation of

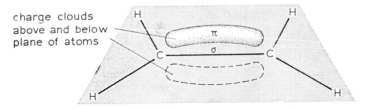

charge clouds above and below plane of atoms

Fig. 10.1

one carbon atom relative to the other is not possible unless a considerable amount of energy is expended to decrease the degree of overlap.

This restriction of the rotation of carbon atoms linked by a π-bond is responsible for the existence of different stable compounds called *cis*- and *trans*- (geometrical) isomers. These compounds contain the same atoms and bonds but differ in the geometric position of corresponding atoms. The *cis*- isomer has similar atoms on the same side of the double bonds. The *trans*- isomer has similar atoms on opposite sides of the double bond. 1,2-dichloroethylene, for example, exists as two stable distinct compounds:

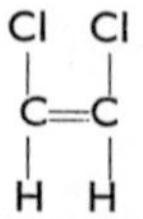

cis-dichloroethylene 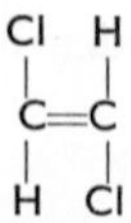   trans-dichloroethylene

A third isomer also exists in which chlorine atoms are attached to the *same* carbon atom:

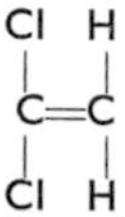

1,1-dichloroethylene

Maleic and fumaric acids

There are very many examples of geometrical isomers known. The difference in structure is well illustrated by the chemistry of maleic and fumaric acids. These acids are represented by the molecular formula $(CH.COOH)_2$, and the isomers are as follows:

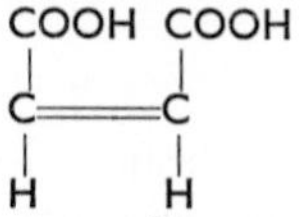

cis- 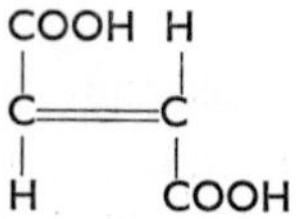   trans-

The *cis*- form has been assigned to maleic acid for the following reasons:

(i) Maleic acid has a much lower melting point (130°C compared to 300°C for fumaric acid), showing that its molecule is the less symmetrical of the two, and so cannot pack so closely in the crystal.

(ii) Maleic acid heated above its melting point readily forms maleic anhydride. Fumaric acid undergoes anhydride formation under more severe conditions.

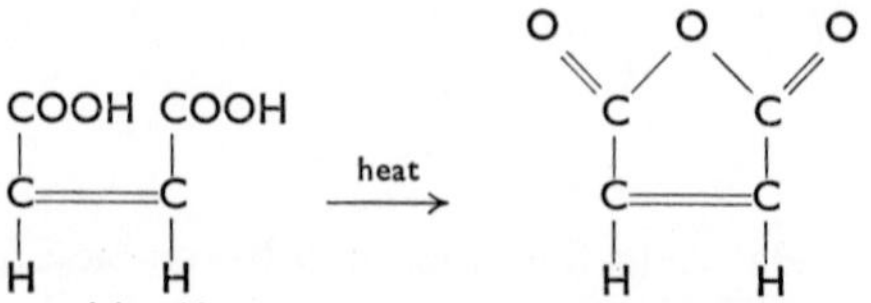

maleic acid   maleic anhydride

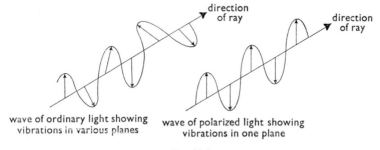

(iii) Maleic anhydride, when hydrolysed by cold water, produces maleic acid only:

Conversion to fumaric acid would necessitate rearrangement which cannot occur at room temperature.

2. Optical isomerism

Before considering optical isomerism it is necessary to understand what is meant by *polarized light*. The vibrations in a light wave are transverse, that is, they take place in all directions perpendicular to the direction of the ray. When light is passed through a crystal of the mineral calcite (a

direction
of ray

direction
of ray

wave of ordinary light showing
vibrations in various planes

wave of polarized light showing
vibrations in one plane

Fig. 10.2

form of calcium carbonate), the emergent beam consists of two types of ray, usually labelled the ordinary and the extraordinary rays. In the ordinary ray the vibrations are in all directions perpendicular to the ray, but in the extraordinary ray the vibrations are all in one plane. Light in this ray is said to be *plane polarized*. Fig. 10.2 shows the vibrations in an ordinary ray and in a plane polarized ray.

A Nicol prism is used for producing a beam of plane polarized light. This consists of a bisected crystal of calcite with its two halves cemented with Canada balsam. When light is passed through this prism the extraordinary ray passes through giving a beam of polarized light, and the ordinary ray is reflected at the interface.

There are many solid substances which rotate the plane of polarized light. Forms of sodium periodate ($NaIO_4$), sodium chlorate ($NaClO_3$), lactic acid ($C_6H_6O_3$) and tartaric acid ($C_4H_6O_6$) do this and are said to be *optically active*. If a beam of polarized light from one Nicol prism (the polarizer) is passed through a second prism (the analyser) the emergent beam has maximum intensity when the axes of the prisms are aligned. The intensity of the emergent beam is a minimum when the axes of the prisms are at right angles (crossed Nicols). The amount of rotation of polarized light by a substance is determined by placing the substance between polarizer and analyser previously set to give an emergent beam of maximum intensity. The analyser is now rotated about its axis until maximum intensity is again observed. The angle of rotation is read off on a circular scale. This is the principle of an instrument called a polarimeter, which is designed to determine accurately the amount of rotation of polarized light.

Optical activity was first noticed for crystals of some inorganic substances. It was observed that sodium periodate exists in two different crystal forms; one form was found to rotate the plane of polarized light to the right, the other form caused rotation to the left. Examination of the crystal forms shows:

(*a*) the crystals are asymmetric,
(*b*) the crystals are non-superimposable mirror images of each other.

The word asymmetric means lacking certain elements of symmetry, e.g. having no plane of symmetry (i.e. the crystal cannot be cut so that two identical halves are obtained). Crystalline substances which are optically active are said to be *enantiomorphic*.

The organic substances mentioned above are enantiomorphic, but differ from the inorganic compounds in one very important way. If different crystal forms of lactic acid, for example, are dissolved in water the resulting solutions retain the optical activity. It is therefore reasonable to assume (*a*) that the optical activity is a result of molecular structure, not just crystal structure, (*b*) that the molecule of lactic acid is asymmetric, and (*c*) the molecules of the two different forms are non-superimposable mirror images of each other.

The optical isomers of lactic acid have the structures shown in Fig. 10.3. These isomers are called dextro-lactic acid (rotating the plane of polarization to the right) and laevo-lactic acid (rotating the plane of polarization to the left).

It is clear that optical isomerism is very probable if the molecule contains an asymmetrical carbon atom, that is a carbon atom attached to four different groups.

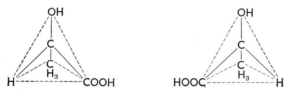

Fig. 10.3

Fischer projections

Representation of a three-dimensional structure by a perspective method is difficult and inconvenient for large molecules. A simple two-dimensional projection (called a *Fischer projection*) is extremely useful. Fig. 10.4 shows the Fischer projection representation of a tetrahedral structure.

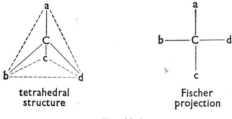

Fig. 10.4

There is one important restriction when manipulating the projections to investigate superimposability; this is that the structure must not be removed from the plane of the paper. It is clear that the forms of lactic

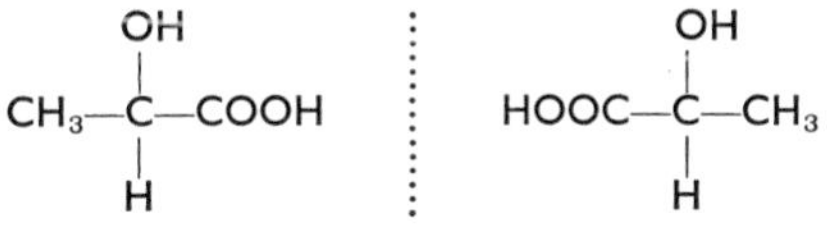

$$CH_3-\overset{\displaystyle OH}{\underset{\displaystyle H}{C}}-COOH \qquad HOOC-\overset{\displaystyle OH}{\underset{\displaystyle H}{C}}-CH_3$$

Fig. 10.5

acid shown by Fischer projections (Fig. 10.5) cannot be superimposed by rotating in the plane of the paper; they are therefore different molecules.

Care must be taken when representing substances not containing an asymmetric carbon atom by Fischer projection. For example, glycollic acid, $CH_2(OH)COOH$, has the structure shown in Fig. 10.6. Examina-

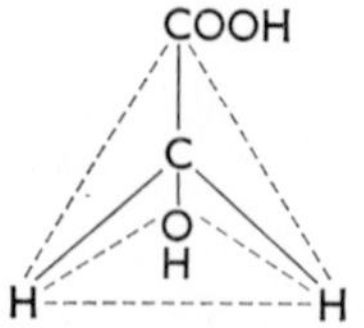

Fig. 10.6

tion shows that it has no non-superimposable mirror-image isomer. Fischer projections (a), (b) and (c) can be drawn (Fig. 10.7). Forms (b)

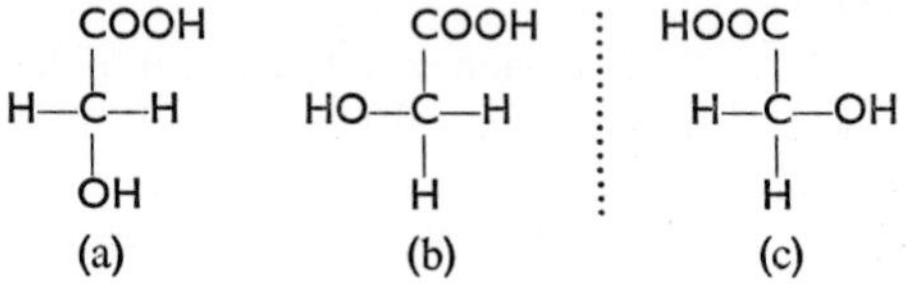

Fig. 10.7

and (c) are apparently mirror-image isomers. Representation by (b) or (c) is wrong because the molecule has been drawn as if it had no plane of symmetry, and so, in two dimensions, it appears to be asymmetric. Form (a) shows clearly the plane of symmetry present in the three-dimensional molecule.

Tartaric acids

Many substances contain more than one asymmetric carbon atom. Tartaric acid,

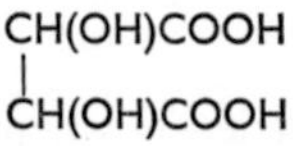

$$CH(OH)COOH$$
$$|$$
$$CH(OH)COOH$$

is a well-known example. For such substances the possibility of several stereoisomers is increased. Using Fischer projections, it is evident that there are three possible molecular arrangements for tartaric acid (Fig. 10.8). Compounds (a) and (b) are mirror-image isomers referred to as

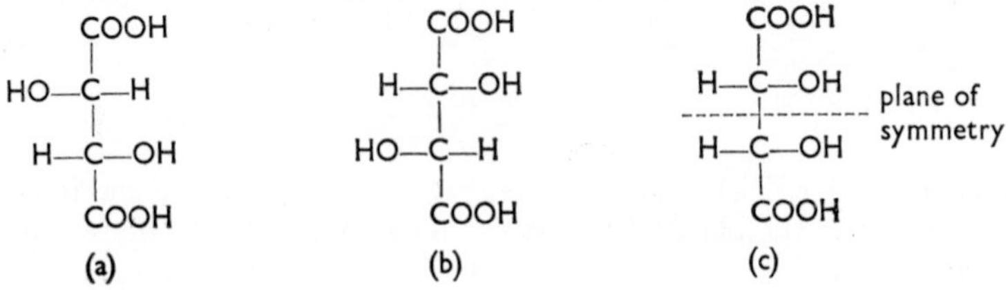

Fig. 10.8

dextro- and laevo-tartaric acids, whilst (c) is not optically active and is called meso-tartaric acid. The three compounds have been isolated and their properties examined. As expected, the d- and l- forms show marked similarity, their melting points, solubilities and ionization constants being identical. The essential difference is that the d- form rotates the plane of polarized light to the right and the l- form rotates it an equal amount to the left. Meso-tartaric acid, in contrast, has a lower melting point, it is less soluble, and its ionization constants are greater than for the d- and l- acids.

Stereochemistry of cyclic compounds
The cyclic compounds cyclopropane dicarboxylic acid and cyclohexane dicarboxylic acid are interesting because they show both geometrical and optical isomerism. Cyclopropane dicarboxylic acid can exist in three isomeric forms (Fig. 10.9). The carbon atoms in the ring are in one

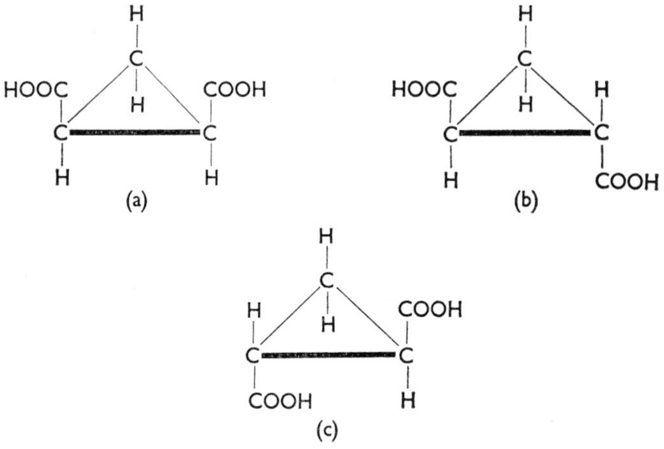

Fig. 10.9

plane and the attached groups are above and below each carbon atom. The three forms of cyclopropane dicarboxylic acid are then, (a) *cis*-form, (b) the d-*trans*-form, and (c) the l-*trans*-form. Compounds (b) and (c) are optical isomers. There are eight isomers of cyclohexane dicarboxylic acid. Four of these are shown in Fig. 10.10. As before, carbon atoms in the ring are in the same plane and other atoms or groups are attached above and below these carbon atoms.

The reader should attempt to find for himself the names and formulae of the other four isomers. It will be seen that there are three pairs of *cis*-, *trans*- isomers and that two of the *trans*- isomers have optically active d- and l- forms.

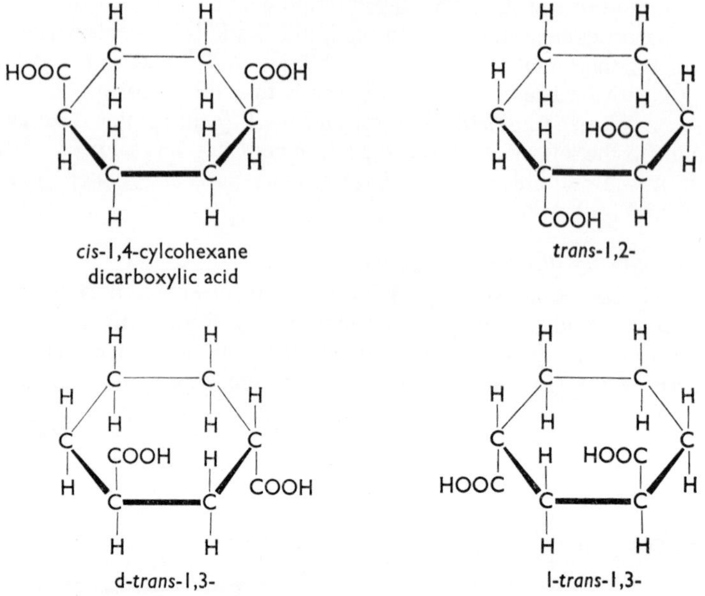

cis-1,4-cylcohexane
dicarboxylic acid

trans-1,2-

d-trans-1,3-

l-trans-1,3-

Fig. 10.10

Racemic mixtures

Synthesis of an optically active compound such as lactic acid will produce a mixture of both the d- and l- isomers in equal amounts. This mixture is called a *racemic mixture* or *racemate*. The solid contains d- and l- molecules present in equimolecular proportions and is often referred to as the dl- compound. It is optically inactive and has different physical properties from the d- and l- forms. Tartaric acid, then, can exist in four different forms, two optically active (the d- and l- acids) and two optically inactive (the racemic and the meso-acids). The meso-form is said to be internally compensated, while the racemate is said to be externally compensated.

Resolution of racemic mixtures

The process of separating a racemate into its optically active forms is called *optical resolution*. Several methods of resolution have been used; these include:

(i) *Formation of diastereoisomers*   When a racemic mixture is made to combine with an optical isomer of a second compound two optically active compounds are produced. These compounds although optically active are not mirror images of one another and are referred to as diastereoisomers. They have different physical properties, in particular different solubilities, so that separation by fractional crystallization can be effected.

If dl-lactic acid, for example, is neutralized by the optically active base d-strychnine the diastereoisomers formed are the salts d-strychnine l-lactate and d-strychnine d-lactate. Representation of the compounds by symbols, Fig. 10.11, shows that these salts are not *mirror-image* isomers.

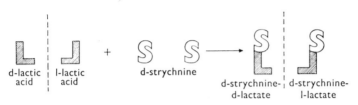

Fig. 10.11

These salts are separated by fractional crystallization. Treatment with caustic soda will remove the optically active base. Acidification with mineral acid will then release the optically active acid.

(ii) *Use of micro-organisms* The chemical processes taking place in living organisms frequently show preferential selection of one of the isomers in a dl mixture. Certain moulds, for example, grown in salt solutions containing racemic acid will consume the d-form much more rapidly than the l-form. After a time isolation of the l-form can be effected. The use of micro-organisms for actual resolution of dl mixtures is difficult and rarely attempted. The method can be utilized, however, in detecting whether or not a given material is actually a dl mixture, since an optically inactive solution will gradually develop optical activity.

(iii) *Separation by hand* Several crystalline substances have been resolved into their d- and l-forms by carefully picking out the 'left-handed' from the 'right-handed' crystals. The method is lengthy and tedious and requires a good deal of skill. It does not compare at all favourably with method (i), which is clearly the best method of resolution.

Recent work in stereochemistry

In the last decade considerable advances have been made in the investigation of the molecular structure of organic substances. The use of infra-red (IR) and nuclear magnetic resonance (NMR) spectra has resulted in rapid progress in understanding the problem of organic structure. Nuclear magnetic resonance spectra show how the functional groups are arranged in the molecular framework, whilst infra-red afford evidence of the various functional groups present. There are two important fields where exciting progress has been made; these are conformational analysis and chemical topology.

1. Conformational analysis

It has been shown that rotation about a single carbon–carbon bond is not completely free as was once supposed. The internal energy change in the ethane molecule as one methyl group rotates relative to another is known to vary between 0 and $11 \cdot 7 \text{ kJ mol}^{-1}$. Forms of ethane associated with these extremes are shown by Newman projections (Fig. 10.12) in

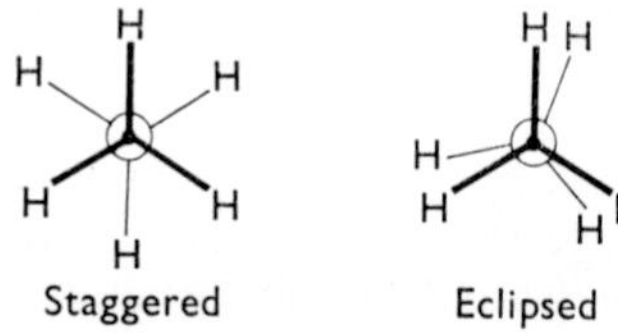

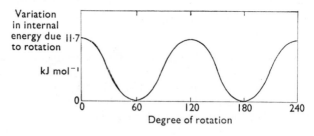

Staggered   Eclipsed

Fig. 10.12

which the two carbon atoms (one behind the other) are represented by ⊙. The degree of rotation is zero in the eclipsed arrangement and 60° in the staggered arrangement. The variation in the internal energy of the ethane molecule with rotation may be illustrated graphically as in Fig. 10.13.

Fig. 10.13

Molecular forms which differ because of different degrees of rotation about a single bond are called *conformational isomers* or *conformers*. It must be emphasized that conformers are usually incapable of independent existence, but exist in equilibrium mixtures.

The ethane molecule oscillates about the minimum energy (staggered) position but will often acquire sufficient energy to overcome the energy barrier and pass through the eclipsed state.

The study of the relationship between the properties of a compound and its conformation is called *conformational analysis*. It is not fully understood why there is an energy barrier to rotation, but in more complicated molecules it has been shown that repulsive forces between attached groups play an important part. The conformations for ethylene dichloride and the corresponding internal energy changes are shown in Fig. 10.14. The energy is a maximum when the bulky chlorine atoms

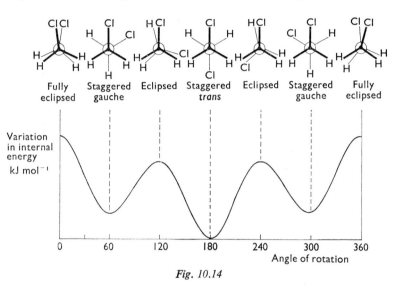

Fig. 10.14

are as close as possible and a minimum when they are as distant as possible. It is apparent that the molecule favours the staggered *trans*-conformation. The energy curve also shows that there is less repulsion between groups when hydrogen and chlorine eclipse than when chlorine eclipses chlorine.

The alternative *boat* and *chair* conformations of cyclohexane, shown in Fig. 10.15, were recognized more than seventy years ago. The mole-

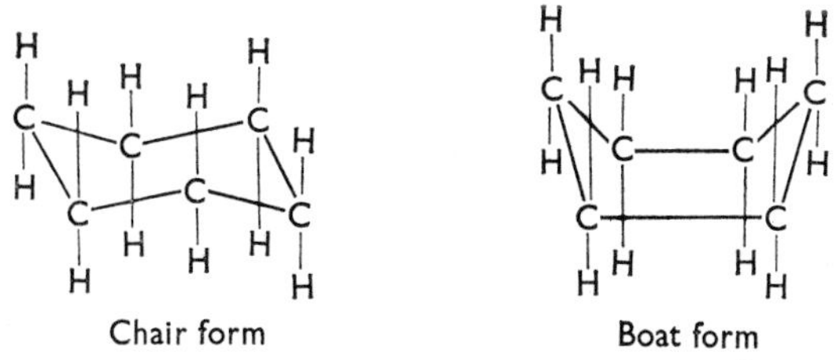

Fig. 10.15   Cyclohexane

cule has the chair form most of the time, this being the more stable of the two forms since it contains fewer eclipsed groups. Neither the boat nor the chair forms are rigid structures; for example, the prow of the boat moves from carbon atom to carbon atom around the ring, as shown in Fig. 10.16.

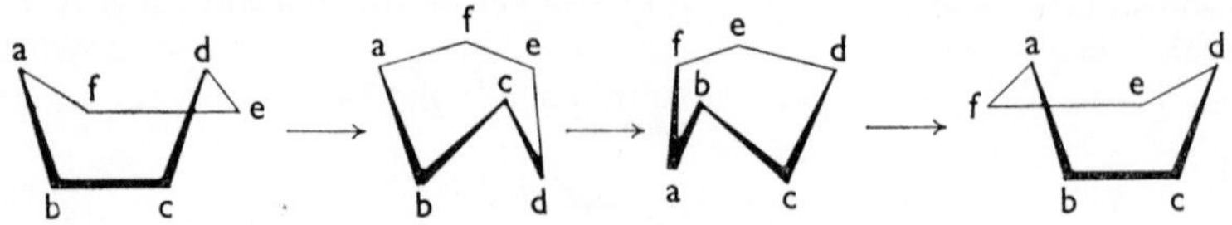

Fig. 10.16

One of the earliest examples of restricted rotation was the discovery of the optical isomers of 2,2′-dinitro-6,6′-dicarboxydiphenyl (Fig. 10.17).

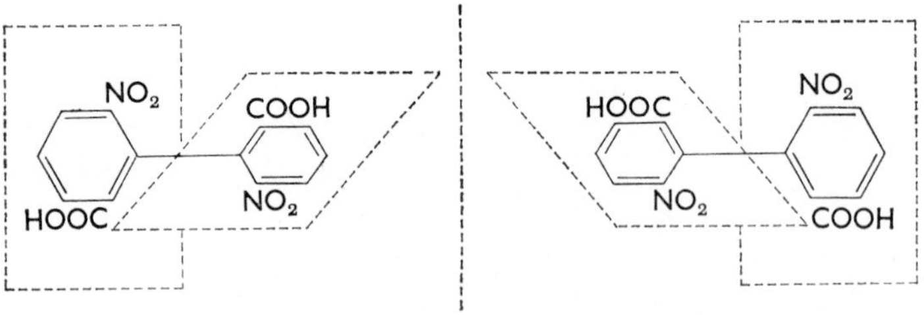

Fig. 10.17   Isomers of 2,2′-dinitro-6,6′-dicarboxydiphenyl

The bulky substituted groups probably prevent free rotation in this case. It is interesting to note that these non-planar optical isomers are asymmetric but do not contain an asymmetric atom.

2. Chemical topology

This is the study of the various structures possible in closed-chain (ring) compounds, when the rings are linked, twisted or knotted. A structure made up of a pair of ring molecules linked only by threading is isomeric with a pair of unlinked molecules. Such compounds are called *topological isomers*.

The possibility of making linked rings was first considered over fifty years ago. Real progress in this kind of synthesis, however, did not begin until 1947 when compounds comprising rings large enough for threading were first made in sufficient quantities. Work with scaled models showed that a molecule would require 20 carbon atoms in the ring if threading were to be feasible. Large rings, of course, can be threaded more easily but they are harder to obtain because their small solubility makes preparation and isolation difficult. The successful synthesis of a catenane (two threaded rings) was carried out with an optimum 34-carbon ring. Synthesis simply involves making a 34-carbon ring; statistically, a few per cent of the rings can be expected to interlock.

Louis Barasch and Edel Wasserman, working at the Bell Laboratories, recently prepared and identified a 34,34-catenane. This compound was made with two different interlocking rings A and B. The B ring contains two oxygen atoms and has the molecular formula $C_{34}H_{65}O_2$. The A ring is a hydrocarbon of molecular formula $C_{34}H_{68}$ in which five of the hydrogen atoms are deuterium* (heavy hydrogen). The catenane is represented diagramatically in Fig. 10.18. Separation of locked and

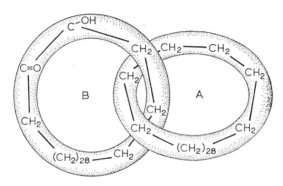

Fig. 10.18

unlocked rings was effected by column chromatography. Identification of the catenane containing deuterium is possible by infra-red spectroscopy.

Although experiments indicate clearly that a catenane was formed, so far only a few milligrammes have been isolated. When sufficient material is available for the determination of molecular weight it will be clear whether the substance contains single or double rings.

The door to this exciting field of molecular structure is open and chemists are eagerly speculating as to the possibility of more complex molecules. When the number of carbon atoms in the ring is increased there is a greater possibility of topological isomers. Some of the possible structures are indicated in Fig. 10.19. It can be seen that structures (a) and (b) are left-hand and right-hand knots and would be mirror-image isomers. Rotation of polarized light therefore could be used for identifying such knotted rings.

Topological isomers may well be obtained eventually from double-stranded ring molecules. These molecules consist of two rings of atoms

* Deuterium is used only as an IR label; the compound could be made using ordinary hydrogen.

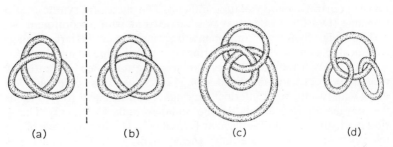

Fig. 10.19

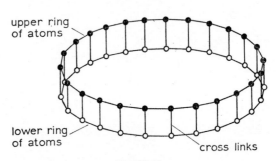

Fig. 10.20

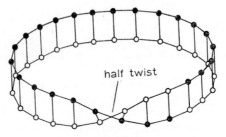

Fig. 10.21

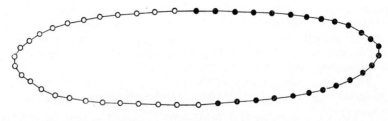

Fig. 10.22

which are cross-linked as indicated in Fig. 10.20. Suppose a double-ring structure contains a half twist as shown in Fig. 10.21. When the cross-links in such a molecule are severed a larger single-ring molecule is obtained as shown in Fig. 10.22.

A double ring containing a full twist severed in the same way would produce a pair of interlocking rings. Separation of a double ring with three half-twists will give a single-knotted ring. The reader can best investigate these and other possibilities using a loop of paper which can be cut through parallel to the edges with a pair of scissors as indicated in Fig. 10.23.

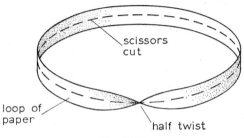

Fig. 10.23

A double-stranded ring molecule containing atoms of silicon instead of carbon in the chain has already been prepared. Further progress may eventually lead to the formation of isomers in the manner indicated above.

Large single-ring molecules have actually been found in living organisms. Giant molecules of this type make knotting easy, but so far investigation has not indicated that topological isomers exist in nature.

In conclusion it can be said that whereas much of chemical topology is speculation at the moment, a breakthrough has been made. Perhaps work in this field during the next decade will result in the preparation of many more unusual molecules.

Isomerism in inorganic compounds

Werner's theory

The formation of complex (coordination) compounds by the transition elements was dealt with in Chapter 9. There it was shown that the incomplete inner shells in the transition elements provide a means of bonding in which unpaired electrons in atoms or groups fill in the vacant orbitals in the metal or metal ion. Developments which led to modern theories of coordinate bonding began with a comprehensive theory proposed by

8+

Alfred Werner in 1893. The postulates of Werner's theory may be summarized thus:

(i) Metals possess two types of valency, primary (ionic) and secondary (non-ionic or coordinate).

(ii) Metals have a fixed number of secondary valencies, either 4 or 6 (others are recognized today). The number of secondary valencies is called the *coordination number* of the metal.

(iii) Primary valencies are satisfied by negative ions, whereas secondary valencies are satisfied by either negative ions or neutral molecules. The number of coordinated groups must equal the coordination number of the metal.

(iv) The secondary valencies are directed in space around the central metal atom or ion. For a coordination number of 6 the valencies are directed toward the apices of a regular octahedron, whilst a coordination number of 4 requires either a planar or a tetrahedral distribution. Werner predicted that the spatial arrangements of complexes should give rise to several kinds of isomeric compound.

Table 10.1, showing some of the platinum-ammines in which the coordination number is six, illustrates Werner's formulation of complex

$[Pt(NH_3)_6]Cl_4$	Hexamminoplatinum (IV) chloride
$[Pt(NH_3)_5Cl]Cl_3$	Chloropentamminoplatinum (IV) chloride
$[Pt(NH_3)_4Cl_2]Cl_2$	Dichlorotetramminoplatinum (IV) chloride
$[Pt(NH_3)_3Cl_3]Cl$	Trichlorotriamminoplatinum (IV) chloride

Table 10.1

compounds with their modern names. The complex ion (enclosed in the square brackets) takes part in chemical reactions as though it were a single radical. Therefore, the proportion of ionizable chlorine in the compounds varies from four parts in four to one part in four.

Werner's theory explained satisfactorily the nature and properties of many complex compounds. Subsequent work by Werner and others substantiated the theory, though it was not until twenty years later that Werner succeeded in isolating the inorganic optical isomers he had predicted.

Later developments in the theory of coordination bonding, culminating in the electrostatic field (ligand field) theory (see p. 182) have resulted in considerable progress in explaining the *nature* of the bonding in complexes. The theory does not conflict with the ideas of Werner, who deserves lasting credit for providing the conceptual framework on which to build.

Ligands

The central metal atom (or ion), with the attached groups or ligands, is referred to as the coordination sphere, and is shown enclosed in square brackets, thus: $[Fe(CN)_6]^{3+}$. Ligands can occupy one (unidentate), two (bidentate), three (tridentate) or more positions in the coordination sphere. Groups occupying more than one position (polydentate groups) are known as *chelates* (from the Greek *chele*, a crab's claw). Table 10.2 shows some typical ligands.

Unidentate	H_2O, NH_3, CN^-, Cl^-, Br^-, I^-, NO_2^-
Bidentate	$C_2O_4^{2-}$, CO_3^{2-}, SO_4^{2-}, SO_3^{2-}, $NH_2CH_2CH_2NH_2$*
Tridentate	$NH_2CH_2CHNH_2CH_2NH_2$†

* Ethylenediamine (en)   † Triaminopropane (tp)

Table 10.2

Several types of isomerism, including stereoisomerism, were predicted by Werner. A discussion of these types and some of the experimental evidence supporting the theory is outlined below.

Stereoisomerism

The three spatial arrangements according to Werner for the coordination sphere are shown in Fig. 10.24. For each of these configurations

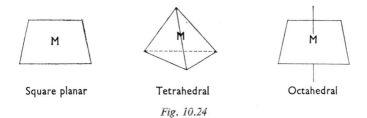

Square planar   Tetrahedral   Octahedral

Fig. 10.24

isomers should be possible. Optical isomers for the tetrahedral and octahedral arrangements are shown in Fig. 10.25, where a, b, c etc., represent unidentate ligands and BD represents a bidentate ligand.

Geometrical (*cis*-, *trans*-) isomers should exist for the square planar and octahedral arrangements, as indicated in Fig. 10.26. Compounds corresponding in formulae to these and other possible structures have been shown to exist. Before discussing examples, it is worth noting the

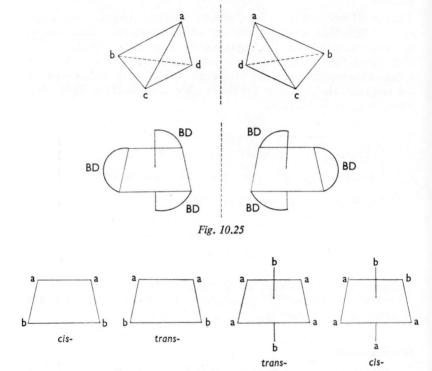

Fig. 10.25

Fig. 10.26

current convention for naming complexes. This may be summarized by the following rules:

(1) The ligands are named first followed by the central atom.
(2) The ligands are named in the following order:

    (i) simple anionic ligands: H^-, O^{2-}, OH^-, S^{2-}, I^-, Br^-, Cl^-, F^-,
   (ii) other inorganic anions,
  (iii) organic anions,
   (iv) H_2O, NH_3,
    (v) other inorganic neutral ligands,
   (vi) neutral organic ligands.

(3) The names of anionic ligands end in -o, while neutral ligands use the name of the molecule, except for water, named aquo, and ammonia, named ammine.
(4) The number of simple ligands is indicated by a prefix; e.g. di-, tri-, and so on. For more complicated ligands the prefixes bis-, tris-, and so on, are used.

(5) In anionic complexes, the name of the central atom ends in -ate, but in neutral or cationic complexes the name of the element is used.
(6) The oxidation number of the central atom is indicated by a Roman numeral in brackets after the name of the atom.

Two compounds of formula $CoCl_3.4NH_3$ are known to exist; one compound is violet and the other is green. Both compounds contain only two ions, and quantitative precipitation with silver nitrate solution shows a single chloride ion present in each substance. They are, in fact, geometrical isomers, as shown in Fig. 10.27.

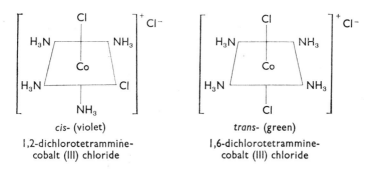

cis- (violet)
1,2-dichlorotetrammine-
cobalt (III) chloride

trans- (green)
1,6-dichlorotetrammine-
cobalt (III) chloride

Fig. 10.27

Another well-known example of geometrical isomerism is that of dinitrotetrammine cobalt (III) chloride. The *flaveo* (golden-yellow) and *croceo* (crocus yellow) forms have similar structures to those in Fig. 10.27.

Square planar geometrical isomers have also been isolated. The substance $Pt(NH_3)_2Cl_2$ exists in two forms; both are yellow, but the *trans*-isomer is deeper in colour. These forms are shown in Fig. 10.28.

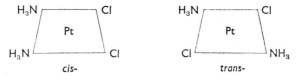

cis-   trans-

Fig. 10.28   Dichlorodiammineplatinum (II)

Numerous optically active isomers have been isolated. Tris(ethylene-diammine) cobalt (III) ions exist as mirror-image isomers, as shown in Fig. 10.29. Other examples of optically active ions are

$$[Cr(en)_3]^{3+}, \quad [Zn(en)_2]^{2+}, \quad [Co(C_2O_4)_3]^{3-} \quad \text{and} \quad [Cr(C_2O_4)_3]^{3-}$$

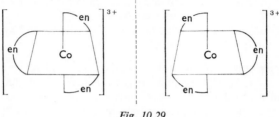

Fig. 10.29

The dichlorobis(ethylenediamine) cobalt (III) ion exists in *cis-* and *trans-* forms. The *cis-* form is asymmetric, and mirror-image isomers have been isolated. These structures are shown in Fig. 10.30.

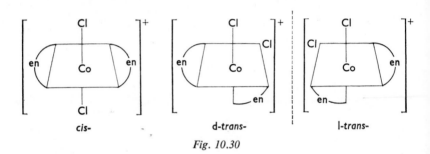

Fig. 10.30

Ionization isomerism

Compounds which have the same molecular formula but which give rise to different ions in solution are called ionization isomers. The compound $[Co(NH_3)_5Br]SO_4$ is purple and in solution will yield sulphate ions. It is isomeric with the red compound $[Co(NH_3)_5SO_4]Br$ which gives bromide ions in solution. Other examples are

$$[Pt(NH_3)_4Cl_2]Br_2 \quad \text{and} \quad [Pt(NH_3)_4Br_2]Cl_2$$
$$[Co(NH_3)_5NO_3]SO_4 \quad \text{and} \quad [Co(NH_3)_5SO_4]NO_3$$

Hydrate isomerism

Water may coordinate, like ammonia, with a central metal ion or it may be loosely associated with metal ions in lattice positions. Hydrate isomerism depends on the way in which water is combined in a substance. A well-known example of hydrate isomerism is shown by hydrated chromium (III) chloride, $CrCl_3 . 6H_2O$. This exists in three isomeric forms,

$$[Cr(H_2O)_6]Cl_3 \text{ (violet)}$$
$$[Cr(H_2O)_5Cl]Cl_2 . H_2O \text{ (green)} \quad \text{and} \quad [Cr(H_2O)_4Cl_2]Cl . 2H_2O \text{ (green)}$$

By precipitation with silver nitrate solution these compounds have been shown to contain 100, $66\frac{2}{3}$ and $33\frac{1}{3}$ per cent ionizable chloride, respectively. Differences in colour, physical and chemical properties are also observed in other examples, such as

$$[Co(NH_3)_3(H_2O)_2Cl]Br_2 \quad \text{and} \quad [Co(NH_3)_3(H_2O)(Cl)Br]Br.H_2O$$

Coordination isomerism

Polynuclear complexes are those which contain more than one central metal ion. If the same kind and number of ligands are distributed in two different ways between the metal ions coordination isomers are obtained. Examples are:

$$[Cu(NH_3)_4]^{2+}[PtCl_4]^{2-} \quad \text{and} \quad [Pt(NH_3)_4]^{2+}[CuCl_4]^{2-}$$

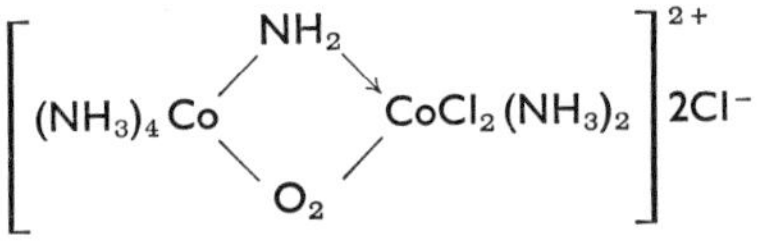

and

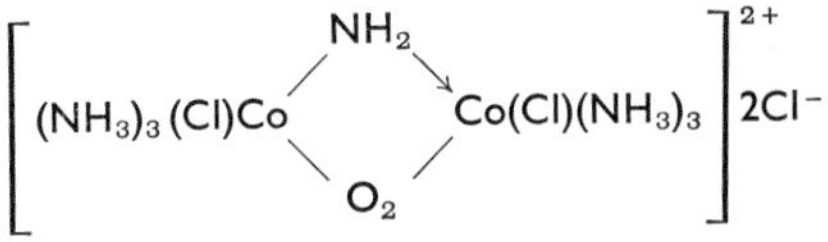

References

Books

Introduction to Organic Chemistry, L. F. Fieser and M. Fieser (D. C. Heath & Co.)

Structure of Organic Molecules, N. L. Allinger and J. Allinger (*Foundation of Modern Organic Chemistry* series) (Prentice-Hall Inc.)

Inorganic Chemistry: An Advanced Textbook, T. Moeller (Wiley)

Collected Readings in Inorganic Chemistry, G. Watt and W. Kieffer (reprint from *J. Chem. Ed.*, 1957–61)

Papers

'Chemical Topology', Edel Wasserman (reprint from *Scientific American*, November 1962)

'A Fragment of Stereochemistry', G. Baddeley (*Education in Chemistry*, **1**, No. 3, July 1964)

'A Summer Short Course in Co-ordination Chemistry', G. Schlessinger (reprint from *Chemistry*, **39**, June 1966, July 1966, August 1966)

Films and Filmstrips

Optical Isomerism. (Filmstrip), Royal Inst. Chem. Series CGA800

11. Application of valency theory to organic chemistry

Valency theory can be used to elucidate the mechanism by which reactions proceed, to explain the properties of functional groups and, most important, it gives the organic chemist a tool for prediction. The last is a possibility which until recent years has been largely overshadowed by the empirical aspects of the subject.

Charge distribution in organic molecules

The key to the problem of applying valency theory successfully to the study of organic chemistry lies in the understanding of the charge distribution in organic molecules. Variation in electron density is the controlling factor in the behaviour of such molecules.

In Chapter 4 it was shown that covalent molecules possessed some ionic character by virtue of the difference in electronegativities of the elements present in the molecule. The reasons for representing gaseous hydrogen chloride as $^{\delta+}H\text{--}Cl^{\delta-}$ have already been discussed. It is clear that non-uniform charge distribution will occur in many organic molecules.

The charged centres produced by the non-uniform charge distribution are attacked by two important kinds of reagent: nucleophiles and electrophiles. Electron-donating groups (Lewis bases) are known as *nucleophilic reagents* or *nucleophiles*. They will attack a molecule in the position of low electron density. Typical nucleophiles are the bromide ion Br^- and the cyanide ion CN^-. Electron-accepting groups (Lewis acids) are called *electrophilic reagents* or *electrophiles*. These groups attack the molecule in a position of high electron density. The nitronium ion NO_2^+, and the acylium ion $R\text{--}C^+{=}O$, are examples of electrophiles.

There are several important concepts used in attempting to predict the variation in electron density in an organic molecule. These current theories are discussed below and supported where possible with some experimental evidence.

Resonance or mesomerism

It was pointed out in Chapter 3 that it was possible to write two or more formal structures to represent certain compounds (i) without altering the relative positions of nuclei, and (ii) while still fulfilling electron pairing requirements of atoms present in the molecule. In such a case the structure is best represented as a resonance hybrid of the various possible structures. The hybrid structure is a form which has lower energy than any of the formal structures, but which is dependent on the contributions of the various possible arrangements. It must be emphasized that the resonance concept does not imply oscillation of bonds as visualized by Kekulé in his alternative structures for benzene (see below). Neither does it mean that there is a dynamic equilibrium between alternative structures due to movement of atoms (a process known as *tautomerism*). This would mean that more than one kind of molecule was present, whereas in resonance there is only one kind of molecule present. A purple mark produced by overprinting blue with red is a good visual analogy to the resonance concept. The red and blue correspond to the formal structures, and the purple corresponds to the hybrid structure. Pictorial representation of the structure of a hybrid molecule is difficult and inadequate, but the resonance concept serves a useful purpose. It can show the most probable electron density variation in the molecule and also the less probable structures that can arise.

Kekulé visualized the benzene molecule as existing in two forms in

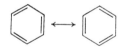

Fig. 11.1

oscillation (Fig. 11.1). Dewar indicated that there were alternatives to the Kekulé structures as shown in Fig. 11.2. Modern theory proposes

Fig. 11.2

that the molecule exists as a resonance hybrid which depends to a varying extent on all the possible formal structures indicated by Kekulé and Dewar. The hybrid molecule is sometimes represented as follows:

The bonds between each carbon atom are assumed to be the same, and neither a simple C–C (σ-bond) nor a C=C (involving a σ- and a π-bond) is present. Experimental evidence from bond length measurement by X-ray or electron diffraction supports this, showing only one carbon to carbon bond length, viz., 0·139 nm, which is intermediate between the bond length of C–C, 0·154 nm, and that of C=C, 0·134 nm.

A second example illustrating the application of the resonance concept is seen in assigning a molecular structure to acetic acid. There are two extreme formulae which can be written as (a) and (b) below.

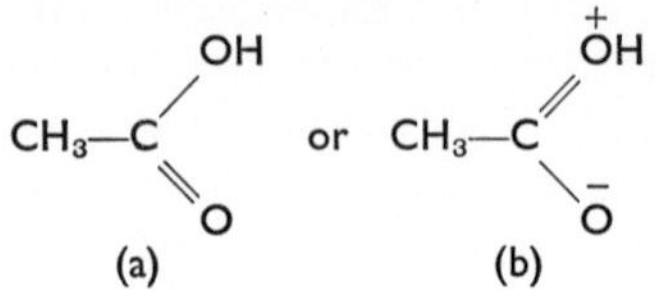

(a)   or   (b)

The structure of acetic acid is more accurately represented as a resonance hybrid, thus:

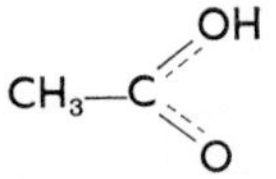

The hybrid molecule has a lower energy state than either of the extremes above. There is experimental evidence which supports this view. X-ray diffraction indicates that the bond lengths in the carboxyl group do not correspond to the expected values of 0·122 nm for C=O and 0·143 nm for C–O, but lie somewhere between these values. The absence of carbonyl activity in carboxylic acids also indicates the absence of a simple C=O group.

The hybrid structures for benzene and for acetic acid (shown above) are attempts at representing structure by utilizing the valency bond approach. The dotted lines are intended to convey delocalized π-bonds as visualized in the molecular orbital model. It was indicated on p. 48 that the benzene molecule could be represented in terms of molecular orbitals by the structure on the right in Fig. 11.3, where the negative

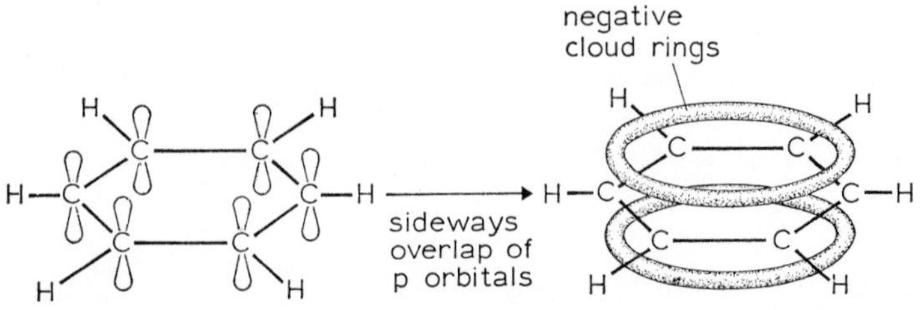

sideways overlap of p orbitals

negative cloud rings

Fig. 11.3

rings above and below the plane of the ring are referred to as a delo-calized π-bond as opposed to localized π-bonds present in ethylene and acetylene (p. 44). Ingold was the first to recognize that certain molecu-lar structures could be more accurately represented as a hybrid of two or more formal structures. He used the word *mesomerism* to describe the hybrid nature of the molecule. Structures can be represented con-ventionally, the probable shift of electrons required to give the hybrid molecule being shown by curved arrows. In a molecule containing a delocalized π-bond, such as a benzene derivative, electron flow into the ring can occur because of the low electron density of the bonds. The electron shift in such molecules is frequently referred to as the *mesomeric effect*. It is worth noting that the effect results in a *permanent* state of the molecule. Consider the electron flow into the delocalized π-bonds of phenol. This mesomeric effect can be represented thus:

(i)

(ii)

(i) and (ii) are possible structures, but phenol is best represented as a hybrid structure, dependent to some extent on all the possible forms. Bond length measurement again indicates the absence of a simple C–O bond. Further experimental evidence supporting the proposition that electrons flow into the ring is the fact that phenol is much more readily attacked in the nucleus by electron-seeking groups (electrophiles), such as the nitronium ion, than is the parent hydrocarbon.

Inductive effect

This refers to the polarity produced in a molecule as a result of the higher electronegativity of one atom compared to another. It results in a *permanent* state of the molecule, as does the mesomeric effect. The car-bon–hydrogen bond is used as a standard, and zero effect is assumed in this case. Atoms or groups which lose electrons toward a carbon atom are sometimes referred to as having a $+1$ effect. Such groups will be referred to in the text as *electron-releasing*. Those atoms or groups which draw electrons away from a carbon atom (-1 effect) will be referred to as *electron-attracting*. Table 11.1 shows the *relative* inductive effects of some common atoms and groups.

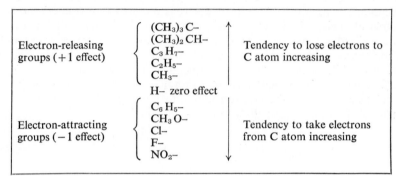

Table 11.1

Evidence for the existence of polarity in many organic molecules and the relative inductive effects due to various groups shown in Table 11.1, comes from a study of dipole moments. The *dipole moment* in a polar molecule is the product of the charge and the distance between opposite charges in the molecule. Values of dipole moment are of the order of 10^{-18} e.s.u. and are usually stated in Debye units (1 Debye unit $= 1 \times 10^{-18}$ e.s.u.). Determination of the dipole moments in a large number of organic compounds makes it possible to calculate approximately the dipole moments due to various groups attached to carbon. It is found experimentally (see carboxylic acids, p. 231) that

(a) the inductive effect is transmitted along a chain of carbon atoms, but diminishes with distance along the chain;

(b) the presence of two or more radicals, with the same directional inductive effect, attached to the same carbon atom, causes greater polarization.

Electromeric effect

This refers to the movement of electrons from one part of a molecule to another *at the moment of attack* by a reagent. (Thus, unlike the inductive and mesomeric effects, it is not a permanent state of the molecule.) The effect is a mutual interaction, polarity being produced in the reagent and the reactant. For example, when ethylene and bromine come into close contact the polarized bromine molecule can interact with the

π electrons in ethylene to form a transitory intermediate compound (π complex) as represented by the structure (a):

$$CH_2{=}CH_2 \quad \rightleftharpoons \quad \begin{array}{c} CH_2{=}CH_2 \\ \vdots \\ Br^{\delta+} \\ | \\ Br^{\delta-} \end{array} \quad (a)$$

$$Br{-}Br$$

Formation of the complex can result in the two step addition of bromine thus:

$$\begin{array}{c} CH_2{=}CH_2 \\ \vdots \\ Br^{\delta+} \\ | \\ Br^{\delta+} \end{array} \xrightarrow{(i)} \begin{array}{c} \overset{+}{C}H_2{-}CH_2 \\ | \\ Br \\ Br^- \end{array} \xrightarrow{(ii)} \begin{array}{c} CH_2{-}CH_2 \\ | \qquad | \\ Br \quad Br \end{array}$$

There are two important pieces of experimental evidence which support the electromeric explanation of the mechanism in this particular reaction: (a) reaction will not occur in the absence of a polar solvent such as water; (b) if some sodium chloride is present then a small amount of ethylene chlorobromide is produced, showing that chloride ions are competing with bromide ions in stage (ii) of the reaction.

A molecule which contains alternate single and double bonds (known as a conjugated system) has the π electrons spread out over the whole molecule (delocalized bonds). When addition occurs in this case, the formation of more than one product is possible thus:

$$CH_2{=}CH{-}CH{=}CH_2 \quad \rightleftharpoons \quad \begin{array}{c} CH_2{=}CH{-}CH{=}CH_2 \\ \vdots \\ Br^{\delta+} \\ | \\ Br^{\delta-} \end{array}$$

$$Br{-}Br$$

$$\begin{array}{c} CH_2{=}CH{-}CH{=}CH_2 \\ \vdots \\ Br^{\delta+} \\ | \\ Br^{\delta-} \end{array} \longrightarrow \begin{array}{c} CH_2{=}CH{-}\overset{+}{C}H{-}CH_2 \\ | \\ Br \\ Br^- \end{array} \text{ or } \begin{array}{c} \overset{+}{C}H_2{-}CH{=}CH{-}CH_2 \\ | \\ Br \\ Br^- \end{array}$$

$$\xrightarrow[\text{addition}]{\text{second}} \begin{array}{c} CH_2{=}CH{-}CH{-}CH_2 \\ | \qquad | \\ Br \quad Br \end{array} \text{ or } \begin{array}{c} CH_2{-}CH{=}CH{-}CH_2 \\ | \qquad\qquad | \\ Br \qquad\quad Br \end{array}$$

1,2–dibromobut–3–ene or 1,4–dibromobut–2–ene

Conclusion

There is ample experimental evidence to support the concepts of resonance (mesomerism) and the inductive and electromeric effects. The usefulness of the theory becomes more apparent in the following sections. The properties of selected compounds and functional groups and the mechanisms of some organic reactions will be considered in more detail, utilizing where appropriate the theory which has been discussed. It must be emphasized that more than one effect may be operating for a particular molecule. Reinforcement occurs when the effects are in the same direction and reduction occurs when the effects are directionally opposite. It is often the case that one effect may be very much greater than another, so that the lesser effect may be ignored altogether.

Carboxylic acids

It has been mentioned (p. 226) that acetic acid is best represented in valency bond terms as a hybrid structure, thus:

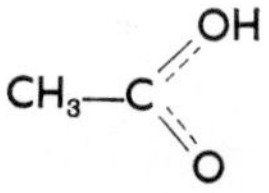

The dotted lines represent a delocalized π-bond. There is an appreciable tendency for electrons to flow into the bond because of its low electron density (mesomeric effect). Electron flow from hydrogen into the delocalized π-bond is strongly reinforced by the inductive effect of the 'double-bonded' oxygen, thus:

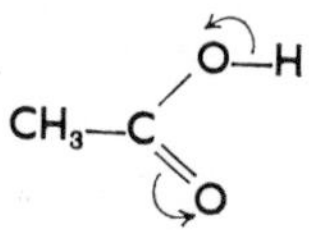

This results in low electron density at the hydrogen atom of the OH group in the acid. Proton loss,

$$CH_3COOH + H_2O \rightleftharpoons CH_3COO^- + H_3O^+$$

by acetic acid should occur fairly readily. Carboxylic acids do in fact show well-defined acidic character. The experimental pK_a value for acetic acid is 4·75.

Carboxylic acids contain a delocalized π-bond, so there is no carbonyl group present. They will not therefore show the nucleophilic addition reactions given by aldehydes and ketones (p. 244). There is, for the same reason, an absence of carbonyl activity in the derivatives of carboxylic acids such as esters and amides. Hybrid structures of esters and amides are shown below (the R indicating an alkyl group):

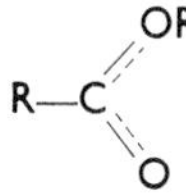

Ester 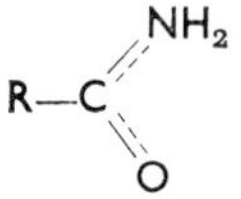   Amide

Inductive effect in acids

(*a*) In an acid R.COOH the tendency for electron shift from the hydrogen atom decreases (and hence acid strength decreases) if R is an electron-releasing group. (The inductive effects of various groups have been given on p. 228.) The expected order of acid strengths would be as shown:

$$C_3H_7COOH < C_2H_5COOH < CH_3COOH < C_6H_5COOH$$
$$\quad 4 \cdot 88 \qquad\qquad 4 \cdot 82 \qquad\qquad 4 \cdot 75 \qquad\qquad 4 \cdot 20$$

Experimental determination of acid strengths (pK_a values are given under each acid) agrees with the expected order.

(*b*) For substituted acids such as $CH_2ClCOOH$ one would expect to find that acid strength had increased (compared to the unsubstituted acid) because of the strong electron-attracting effect of the chlorine atom. Since the inductive effect is cumulative the expected strengths for mono-, di- and trichloroacetic acids would be in the order:

$$CCl_3COOH > CHCl_2COOH > CH_2ClCOOH > CH_3COOH$$
$$\quad 0 \cdot 08 \qquad\qquad 1 \cdot 3 \qquad\qquad 2 \cdot 86 \qquad\qquad 4 \cdot 75$$

The pK_a values given with each acid agree with the expected order.

(*c*) The inductive effect of an atom decreases with the distance of that atom along the chain, so that for α, β and γ chlorobutyric acids, strengths in the order shown would be expected:

$$CH_3CH_2CHClCOOH > CH_3CHClCH_2COOH$$
$$\quad 2 \cdot 86 \qquad\qquad\qquad 4 \cdot 06$$

$$> CH_2ClCH_2CH_2COOH > CH_3CH_2CH_2COOH$$
$$\qquad\qquad 4 \cdot 52 \qquad\qquad\qquad 4 \cdot 82$$

The experimental pK_a values shown, confirm the expected order.

Phenols

Phenol itself is weakly acidic, forming salts with caustic soda. This acidity is due to electron flow from hydrogen to oxygen, an inductive effect reinforced by resonance. The phenoxide ion is stabilized because the negative charge on the oxygen can be shared by the carbon atoms in the ring. The phenoxide ion exists as a resonance hybrid of the structures shown below:

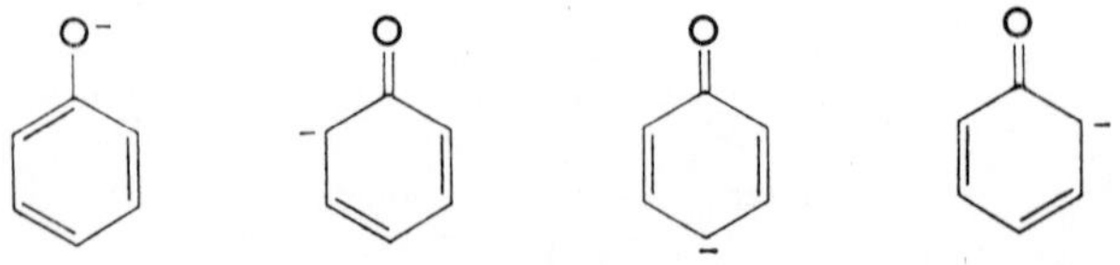

Carboxylic acids, but not phenol, will react with sodium carbonate, liberating CO_2. The greater acidity of carboxylic acids is due to the powerful inductive effect of the 'double-bonded' oxygen in the carboxylic group, already mentioned on p. 230.

An increase in acidity would be expected for o-, m- and p-nitrophenols because electron flow into the ring from the OH group is greatly encouraged by the electron-attracting nitro- groups. A further increase in acidity would be expected for dinitrophenols and for trinitrophenol (picric acid). The expected strengths of phenols is confirmed by pK_a values of 10·0 for phenol, between 7 and 8 for o-, m- and p-nitrophenols, 4·0 for 2,4-dinitrophenol and 0·4 for trinitrophenol.

Alcohols

Alcohols do not contain delocalized π-bonds as do carboxylic acids and phenols. They may be represented by the formula R–O–H. Proton loss would result in formation of the alkoxide ion R–O$^-$, but there is very little tendency for this to occur because the absence of delocalized π-bonds gives no opportunity for stabilization of the ion by resonance. Since alkyl groups are electron-releasing compared to hydrogen there is less electron loss from hydrogen in ROH than from hydrogen in HOH. Alcohols then show no acidity comparable with carboxylic acids and phenols. They are in fact less acidic than water, having pK_a values between 15·5 and 16. A minute degree of acidity is shown by alcohols when they react quietly with sodium or potassium to give hydrogen.

Amines

These nitrogen-containing compounds are bases, like ammonia, by virtue of the lone pair of electrons on the nitrogen atom. They combine with protons thus:

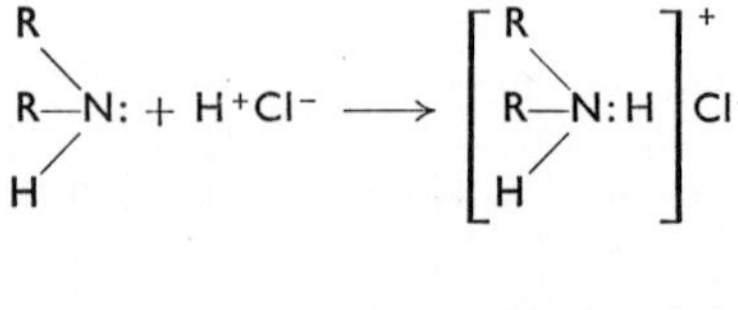

In the case of aliphatic amines, the inductive effect of the alkyl groups (p. 228) would indicate increasing basic nature in the order

$$NH_3 < CH_3NH_2 < C_2H_5NH_2 < C_3H_7NH_2$$

and this is found to be the case.

Aromatic amines are expected to be less basic than ammonia. The benzene ring is electron-deficient and the electron density at nitrogen is decreased because of the tendency for electrons to flow into the delocalized π-bonds (the mesomeric effect, p. 227).

Because of the additive nature of the inductive effect (p. 228), it might be expected that basic character would increase from primary to secondary to teriary aliphatic amines, and decrease from primary to secondary aromatic amines. Fig. 11.4 shows the approximate change in relative basic character determined experimentally for some common amines.

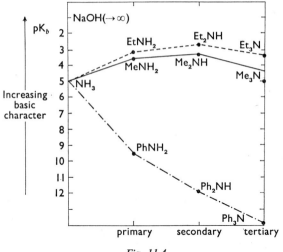

Fig. 11.4

It is noticeable that the increase in basic nature passing from a primary to a secondary aliphatic amine is less than the increase shown in passing from ammonia to a primary amine. There is actually a decrease in basic character going from a secondary to a tertiary amine. This variation is due to the increasing importance of the steric effect of the alkyl groups. It has been shown spectroscopically that the energy required for the lone pair electrons to pass through the nitrogen atom and thus invert the amine molecule is only of the order of 20 kJ mol^{-1}, an amount readily available at room temperature. The position of the alkyl groups and lone pair electrons in an amine will vary between positions (a) and (b) as shown in Fig. 11.5 (where R and R' are alkyl groups).

When there are two, and in particular three, alkyl groups in the molecule the incoming proton has less access to the lone pair than it would if only hydrogen atoms were present. Basic character is, therefore, less than one would expect from consideration of inductive effects only.

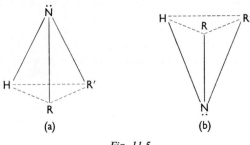

Fig. 11.5

The decrease in basic properties of aromatic compared to aliphatic amines is very noticeable. Aniline, unlike the aliphatic amines, is insoluble in water but does dissolve in dilute acids, giving salt solutions. Diphenylamine on the other hand will dissolve only in concentrated acids.

Reaction mechanisms

Types of reaction

It is useful to distinguish between four basic types of organic reaction: substitution, elimination, addition and rearrangement reactions.

(*a*) *Substitution* reactions involve the replacement of an atom or group in a substance by another atom or group. Alkanes, alkyl halides and alcohols can undergo substitution reactions.

(*b*) *Elimination* reactions refer to the formation of a new substance by the ejection of simple molecules such as water or halogen acid. Alcohols and alkyl halides can undergo this type of reaction also.

(*c*) *Addition* reactions occur when atoms or groups attach themselves to compounds containing double or triple bonds. Alkenes, alkynes and carbonyl compounds readily undergo reaction by addition.

(*d*) *Rearrangement* involves the movement of atoms or groups from one part of a species (molecule or ion) to another. Rearrangement is often one of the steps in a condensation reaction. (A condensation reaction is not a *type of reaction* because it involves more than one step; each step is one of the basic types discussed above.)

Modern theory, to be discussed below, proposes particular mechanisms by which the various types of reaction are thought to proceed. It should be emphasized, however, that the mechanism visualized is only a useful model. It is not intended to represent exactly what is happening. There is experimental evidence to support much of the theory, but ideas are constantly being revised in the light of new knowledge.

Bond breaking and bond making

Organic reactions require the breaking and making of covalent bonds. A molecule XY which is bound by the mutual sharing of a pair of electrons can show bond breaking in the following ways:

$$X:Y \longrightarrow X\cdot + Y\cdot \qquad (a)$$

$$\left. \begin{array}{l} X:Y \longrightarrow X\overline{\cdot} + Y^+ \\ X:Y \longrightarrow X^+ + :Y^- \end{array} \right\} \quad (b)$$

Bond fission in (a) is called *homolysis*, and results in *free radicals*; in (b) the bond fission is called *heterolysis*, and produces *ions*. A large number of organic reactions occur in the presence of ionic reagents such as nitric acid, sulphuric acid and caustic soda; such reactions will occur by an ionic mechanism. Substances which are relatively stable to the usual ionic reagents react by free-radical formation.

(a) SUBSTITUTION REACTIONS

1. Free-radical reactions

The alkanes are frequently described as unreactive compounds. This is true to a large extent if attack by mineral acids, alkalis or oxidizing agents is envisaged. Attack by such reagents requires ionization, and there is little polarity in a C–H bond. Reaction of alkanes takes place by free-radical formation, and when conditions are favourable to this kind of mechanism the alkanes are reactive.

The reaction between chlorine and methane is thought to proceed by a chain reaction. The first step, called the initiation process, involves a chlorine molecule absorbing a quantum of light. The energy absorbed is greater than the bond dissociation energy of the molecule. Separation into atoms occurs thus:

$$\textit{Initiation} \quad Cl_2 + h\nu \longrightarrow Cl\cdot + Cl\cdot$$

The propagating reactions which follow constitute a repeating cycle producing methyl chloride, thus:

$$\textit{Propagation} \quad CH_4 + Cl\cdot \longrightarrow CH_3\cdot + HCl$$

$$CH_3\cdot + Cl_2 \longrightarrow CH_3Cl + Cl\cdot$$

The chain reaction can be halted in three ways, called termination reactions. These are shown below:

$$\text{Termination} \quad Cl\cdot + Cl\cdot \longrightarrow Cl_2$$
$$CH_3\cdot + CH_3\cdot \longrightarrow CH_3CH_3$$
$$CH_3\cdot + Cl\cdot \longrightarrow CH_3Cl$$

The speed at which conversion of an alkane into an alkyl halide occurs depends on the energetics of each of the propagating reactions in the chain. When one of the propagating reactions is endothermic, rate of halogenation at room temperature can be extremely slow. The reaction between bromine and methane is an example: halogenation occurs at a reasonable rate only when heat is applied. Sometimes each of the propagating reactions is highly exothermic, when halogenation is rapid; fluorine, for example, reacts explosively with methane at room temperature.

2. Reactions involving ions (S_N1, S_N2 reactions)

These occur when *nucleophiles* attack electron-deficient sites in the molecule. There are two different processes by which the reaction mechanism can operate; these are labelled S_N1 and S_N2 reactions, S referring to substitution, N to nucleophilic and 1 or 2 indicating first or second order.

S_N1 *reactions*

The mechanism of S_N1 reactions may be written in general terms as occurring in two steps:

(i) $$R_3C\text{—}X \xrightarrow{\text{slow}} R_3C^+ + X^-$$

(ii) $$R_3C^+ + Z^- \xrightarrow{\text{fast}} R_3C\text{—}Z$$
$$\underset{\substack{\text{nucleo-}\\\text{phile}}}{}$$

Here R represents an H atom or alkyl group, X represents a group such as Br, I, OH, etc. and Z represents CN^-, etc.

Step (i) is the slow (rate-determining) step, so that the substitution is first order. The substrate (molecule attacked) only is involved in the rate-determining step. Step (ii) is the very rapid combination of oppositely charged ions.

Characteristics of S_N1 *reactions*

(a) They always occur with the formation of a carbonium ion,

$$\overset{\diagup}{\underset{\diagup}{>}} C^+$$

(b) Since the first step requires ionization, S_N1 reactions are facilitated for a compound R_3CX, (i) if X is an electron-attracting group, and

(ii) if R_3C is an electron-releasing group. (Examples of groups which show electron-releasing and electron-attracting tendencies have been given in the discussion of inductive effect, p. 228.)

(c) Since ionization is important, S_N1 reactions are facilitated if a good ionizing solvent is used.

Geometry of S_N1 reactions

This becomes important when the substrate (molecule attacked) is optically active. Consider then the alkyl halide CabcX (optically active) attacked by the nucleophile OH$^-$ (e.g. a tertiary alkyl halide reacting with sodium hydroxide). In step (i) a carbonium ion is formed. This, as indicated by the electron pair repulsion theory (p. 172), is *planar* (see Fig. 11.6). In step (ii) the nucleophile OH$^-$ can approach either *at the*

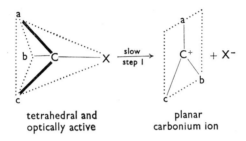

tetrahedral and optically active

planar carbonium ion

Fig. 11.6

front or *at the back* of the planar carbonium ion. When OH$^-$ combines, bonding-pair repulsion causes the molecule to revert to the tetrahedral shape. This results in both laevo and dextro forms of the product

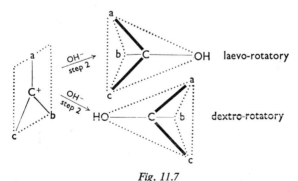

laevo-rotatory

dextro-rotatory

Fig. 11.7

as shown in Fig. 11.7. Hence S_N1 reactions with an optically active substrate generally result in the formation of equal amounts of the isomeric products (racemization). That this is not always the case is explained by preferential attack on one side of the planar carbonium ions. The 'leaving' ion may hinder the approach of the attacking nucleophile, and hence attack on the opposite side of the carbonium ion predominates.

S_N2 reactions

This type of substitution also occurs in two steps. One may represent an S_N2 reaction in general terms thus:

(i)
$$\underset{\text{nucleophile}}{Z^-} + \underset{\text{substrate}}{R_3C\!-\!X} \xrightarrow{\text{slow}} \underset{\text{transition state}}{\overset{\delta-}{Z}\text{---}R_3\overset{}{C}\text{---}\overset{\delta-}{X}}$$

(ii)
$$\overset{\delta-}{Z}\text{---}R_3\overset{}{C}\text{---}\overset{\delta-}{X} \xrightarrow{\text{fast}} Z\!-\!R_3C + X^-$$

In step (i), Z^- approaches and begins to make the bond whilst the X–C bond begins to break. There is a spread of charge and in step (ii) rapid expulsion of the X^- ion occurs. The reaction is second order because the original reactant R_3CX and the nucleophile are both involved in the bond making-breaking, rate-determining step.

Characteristics of S_N2 reactions

(a) No carbonium ion is formed, but a transitory intermediate compound is.
(b) Those groups with minimum inductive effects will favour S_N2 reaction because ionization must not occur.
(c) Since ionization is not required, S_N2 reactions are more probable when non-ionizing solvents such as ether or alcohol are used.

Geometry of S_N2 reactions

Again this is important when the substrate is optically active. Consider the optically active tertiary alkyl halide CabcX attacked by the nucleophile Z^-. The intermediate compound formed, Z^-CabcX, has the groups a, b and c in one plane, perpendicular to the straight line through Z–C–X, radiating out like the spokes of a wheel (Fig. 11.8). When X leaves, bonding pair repulsion causes the molecule to revert to the

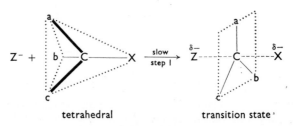

tetrahedral   transition state

Fig. 11.8

tetrahedral shape (Fig. 11.9). There is an inversion of the absolute configuration (the molecule has turned inside out). The change, called a

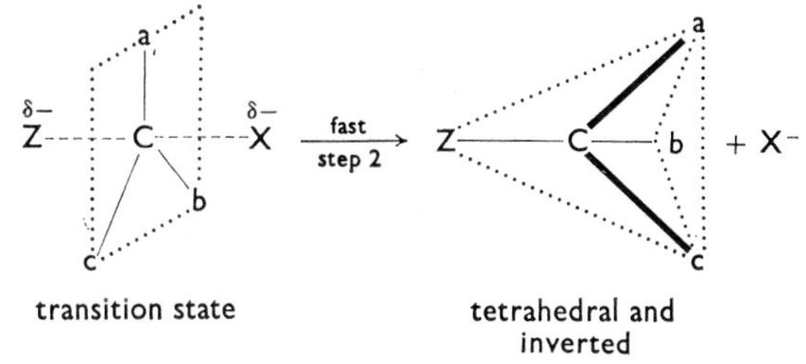

transition state   tetrahedral and
                                                      inverted

Fig. 11.9

Walden inversion, was first noted by Walden in 1896. It does not necessarily imply a change in the direction of rotation of polarized light, though this often occurs.

(*b*) ELIMINATION REACTIONS

E1 reactions

Besides the S_N1 reaction already discussed a carbonium ion can show another type of reaction with a nucleophile. The positively charged carbon atom can cause an electron shift, resulting in increased acidity of the hydrogen atoms to such an extent that a proton can be lost to the attacking base. Thus if alkali reacts with the alkyl halide, $CH_3.CR_2Br$, an alkene is the main product:

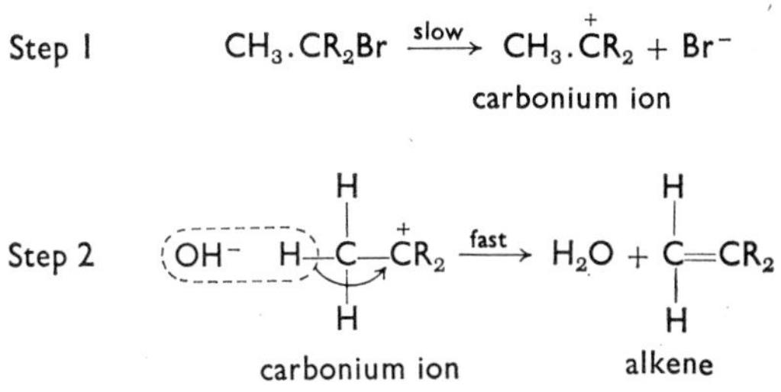

Step 1 $CH_3.CR_2Br \xrightarrow{\text{slow}} CH_3.\overset{+}{C}R_2 + Br^-$
                                        carbonium ion

The reaction is labelled E1, the E standing for elimination (of a molecule of halogen acid) and 1 for first order, since the slow formation of the carbonium ion is the rate-determining step (as explained for S_N1 reactions).

E2 reactions

These are elimination reactions which occur without the previous formation of a carbonium ion. Suppose in the reaction above a non-ionizing solvent was used; this would prevent formation of a carbonium ion. The reaction below is thought to proceed in this instance because electron drift toward the electronegative halogen occurs. The hydrogen atoms become sufficiently acidic to be lost to the attacking base thus:

alkene

The reaction is labelled E2, the elimination of halogen acid in this case being second order since base and halide are both involved in this single-step reaction.

Substitution and elimination reactions of alkyl halides

It has been shown that a compound of the type R_3CX, e.g. a primary, secondary or tertiary alkyl halide, can undergo reaction with an alkali by any one of four different mechanisms. S_N1 and S_N2 reactions will produce an alcohol. E1 and E2 reactions will produce an alkene. (Note: An S_N2 reaction can also produce an ether; this is shown later on p. 242.)

When an alkyl halide and an alkali react, what are the factors which dictate the mechanism by which the reaction proceeds? They are as follows:

(a) *Nature of the alkyl halide*—particularly the inductive effect of the groups present.
(b) *Nature of the solvent used*—S_N1 and E1 reactions require ionization of the reactant, which is facilitated by a good ionizing solvent.
(c) *Concentration of reactants*—the concentration of alkali is important only in the second order S_N2 and E2 reactions.
(d) *Temperature*—high temperatures will favour elimination rather than substitution reactions.

Note: When any alkyl halide reacts with alkali the reaction may occur by more than one mechanism and result in mixed products. However, with a careful control of the conditions mentioned above it is possible to produce a favourable yield of the desired product.

It was pointed out (p. 228) that groups with the highest electron-releasing nature were in the order

$$(CH_3)_3C > (CH_3)_2CH > C_2H_5CH_2 > C_2H_5 > CH_3$$

It is clear then, that carbonium formation occurs most readily with $(CH_3)_3C$ and least readily with CH_3. One might expect tertiary methyl bromide to react with alkali by an S_N1 reaction and the tendency for reaction to proceed by this mechanism to decrease until, perhaps, at ethyl or methyl bromide reaction is by an S_N2 mechanism. Tertiary and secondary alkyl halides in fact do produce alcohols by S_N1 reaction, whereas primary alkyl halides produce alcohols by S_N2 reaction.

There is, however, another important consideration. It was mentioned earlier that aqueous solution favoured S_N1 reactions. It may well be that the use of alcoholic alkali will cause even secondary and tertiary alkyl halides to react by a mechanism other than S_N1. It is found in fact that secondary and tertiary alkyl halides do not react by S_N1 mechanism when alcohol is employed.

A further consideration is the possibility of an elimination reaction occurring. This is favoured (a) when alcohol is the solvent, (b) if the concentration of alkali is high (for E2 mechanism), (c) if the higher alkyl halides are used, (d) if a high temperature is used to make the reaction go.

Table 11.2 shows the main products and the main mechanism when

Conditions \ Alkyl halide	$(CH_3)_3C.Br$	$(CH_3)_2CH.Br$	C_3H_7Br	C_2H_5Br	CH_3Br
Aqueous NaOH	alcohol S_N1	alcohol S_N1+2	alcohol S_N2	alcohol S_N2	alcohol S_N2
Alcoholic NaOH	alkene E2	alkene E2	ether S_N2	ether S_N2	ether S_N2

Table 11.2

aqueous and alcoholic sodium hydroxide react with various alkyl halides.

Examples

(i) Formation of alcohol from ethyl bromide and aqueous alkali (S_N2):

$$OH^- + C_2H_5Br \xrightarrow{slow} \overset{\delta-}{OH}\text{---}C_2H_5\text{---}\overset{\delta-}{Br}$$

$$\overset{\delta-}{OH}\text{---}C_2H_5\text{---}\overset{\delta-}{Br} \xrightarrow{fast} C_2H_5OH + Br^-$$
ethyl alcohol

(ii) Formation of an alkene from isopropyl bromide and alcoholic alkali (E2):

propene

(iii) Formation of ether from methyl iodide and alcoholic alkali (S_N2):

$$C_2H_5O^- + CH_3I \xrightarrow{\text{slow}} C_2H_5\overset{\delta-}{O}\text{------}CH_3\text{------}\overset{\delta-}{I}$$

from

$$\searrow (C_2H_5OH + OH^- \rightleftharpoons C_2H_5O^- + H_2O)$$

$$C_2H_5\overset{\delta-}{O}\text{------}CH_3\text{------}\overset{\delta-}{I} \xrightarrow{\text{fast}} C_2H_5OCH_3 + I^-$$

methyl ethyl
ether

(iv) Formation of alcohol from tertiary butyl bromide and aqueous alkali (S_N1):

$$(CH_3)_3CBr \xrightarrow{\text{slow}} (CH_3)_3C^+ + Br^-$$

carbonium ion

$$(CH_3)_3C^+ + OH^- \xrightarrow{\text{fast}} (CH_3)_3C.OH$$

tertiary butyl
alcohol

Substitution and elimination reactions of alcohols

Alcohols react in the presence of concentrated sulphuric acid to form ethers and alkenes, depending upon the conditions employed. It is not absolutely certain by which mechanism a particular reaction proceeds, but it has been shown that formation of the oxonium ion of the alcohol is a preliminary to reaction of alcohols in the presence of acid, thus:

$$CH_3\text{—}\underset{H}{\overset{H}{C}}\text{—}O\text{—}H + H_3O^+ \longrightarrow CH_3\text{—}\underset{\underset{H}{|}}{\overset{H}{C}}\text{—}\overset{+}{\underset{\underset{H}{|}}{O}}\text{—}H + H_2O$$

from

$$\searrow (H^+ + H_2O \rightarrow H_3O^+)$$

acid

Ether formation

Loss of water by the oxonium ion would produce a carbonium ion which reacts with a further molecule of alcohol (acting as a nucleophile by virtue of lone pairs of electrons on the oxygen atom); i.e. reaction occurs by an S_N1 mechanism, thus:

$$CH_3\text{—}\underset{\underset{H}{|}}{\overset{H}{C}}\text{—}\overset{+}{\underset{\underset{H}{|}}{O}}\text{—}H \xrightarrow{\text{slow}} CH_3\text{—}\overset{H}{\underset{\underset{H}{|}}{C}}{}^+ + H_2O$$

alcohol oxonium ion   carbonium ion

$$CH_3\text{—}\overset{H}{\underset{\underset{H}{|}}{C}}{}^+ + :\underset{\underset{C_2H_5}{|}}{\overset{H}{O}}: \xrightarrow{\text{fast}} CH_3\text{—}\underset{\underset{H}{|}}{\overset{H}{C}}\text{—}\overset{+}{\underset{\underset{C_2H_5}{|}}{O}}$$

nucleophile   new oxonium ion

Proton loss will then produce the ether thus:

$$CH_3-\overset{\overset{\displaystyle H}{|}}{\underset{\underset{\displaystyle C_2H_5}{|}}{C}}-\overset{\overset{\displaystyle H}{|}}{\underset{\underset{\displaystyle C_2H_5}{|}}{O^+}} \longrightarrow CH_3-\overset{\overset{\displaystyle H}{|}}{\underset{\underset{\displaystyle H}{|}}{C}}-O-C_2H_5 + H^+$$

ether

Alkene formation

Again, loss of water by the oxonium ion gives a carbonium ion:

$$CH_3-\overset{\overset{\displaystyle H}{|}}{\underset{\underset{\displaystyle H}{|}}{\overset{+}{C}}}-\overset{}{\underset{\underset{\displaystyle H}{|}}{O}}-H \underset{}{\overset{slow}{\rightleftarrows}} CH_3-\overset{\overset{\displaystyle H}{|}}{\underset{\underset{\displaystyle H}{|}}{C^+}} + H_2O$$

alcohol oxonium ion   carbonium ion

At higher temperature elimination will occur (an E1 reaction; see p. 239). The carbonium ion loses a proton giving an alkene:

$$CH_3-\overset{\overset{\displaystyle H}{|}}{\underset{\underset{\displaystyle H}{|}}{C^+}} + H_2O \overset{fast}{\longrightarrow} CH_2{=}CH_2 + H_3O^+$$

It was once thought that the formation of alkyl hydrogen sulphate was a necessary step in the production of ethers and alkenes from alcohols. If the mechanisms described above are valid we would expect other acids to bring about reaction (since a source of protons seems to be the essential requirement). This is in fact the case as many acids, among them hydrochloric and phosphoric acid, can be used instead of sulphuric acid.

(c) ADDITION REACTIONS

Addition reactions occur readily with those substances possessing localized π-bonds. The important unsaturated compounds showing reactions by addition are alkenes, alkynes, aldehydes and ketones.

Electrophilic addition

This occurs when the unsaturated hydrocarbons mentioned above react with reagents such as Cl_2, Br_2, HCl, HBr, ICl and IBr. (Each of these reagents may be considered to be made up of a strong acid A^+ and a weak base B^-.) The π-bond electrons in these hydrocarbons are more readily available than the electrons in the delocalized π-bond in benzene. This means that positive groups or reagents (electrophiles) can coordinate (using the π electron pair) with the compound to produce a carbonium ion, thus:

$$\underset{\substack{| \\ \text{electrophile}}}{C=C} + A^+ \xrightarrow{\text{slow}} \overset{A}{\underset{| \\ |}{-C-C^+}}$$

This is the slow rate-determining step. The carbonium ion is a very reactive transitory intermediate compound and reacts immediately with the residual negative group of the reagent B^-, thus:

$$\overset{A}{\underset{| \\ |}{-C-C^+}} + B^- \xrightarrow{\text{fast}} \overset{A}{\underset{| \\ |}{-C-}}\overset{B}{\underset{| \\ |}{C-}}$$

A very stable molecule is produced containing σ-bonds in place of the original weaker π-bond. An example of electrophilic addition in this way has already been given in connection with the electromeric effect (p. 228).

Nucleophilic addition

The carbonyl compounds show addition reactions with substances such as HCN, $NaHSO_3$, NH_2NH_2 and NH_2OH. In a carbonyl group the highly electronegative oxygen atom is exerting a strong inductive effect, pulling the π electron pair away from the carbon atom. The electromeric effect caused by the presence of an attacking reagent reinforces the permanent inductive effect so that the electron pair is acquired by the oxygen atom, thus:

$$\underset{|}{C} \overset{..}{\underset{..}{O}} \longrightarrow {}^+\underset{|}{C} - \overset{..}{\underset{..}{O}}: {}^-$$

The carbon atom becomes an electron-deficient site in the molecule and is readily attacked by nucleophilic reagents. The final product usually exists in equilibrium with a small amount of the original carbonyl compound. Nucleophilic reagents attacking the polar molecule are either negative ions such as CN^-, HSO_3^- or molecules with lone pairs such as $:NH_2NH_2$ or $:NH_2OH$.

The reaction between HCN and acetaldehyde may be written:

$$\underset{H}{\overset{CH_3}{\underset{|}{C}}}\overset{..}{\underset{..}{O}} \xrightarrow[\substack{\text{I and E} \\ \text{effects}}]{HCN} {}^+\underset{H}{\overset{CH_3}{\underset{|}{C}}}\overset{..}{\underset{..}{O}}:{}^- \xrightarrow[\substack{\text{nucleophilic} \\ \text{addition}}]{CN^-} \underset{H}{\overset{CH_3}{\underset{|}{NC-C-O^-}}} \xrightarrow{H^+} \underset{H}{\overset{CH_3}{\underset{|}{CN-C-OH}}}$$

The reaction between NH_2OH and acetone may be written:

$$CH_3-C(=O)-CH_3 \xrightarrow[\text{effects}]{\text{I and E}} {}^{+}C(CH_3)(CH_3)-\ddot{O}:^{-} \xrightarrow[\text{addition}]{\text{:}NH_2OH \text{ nucleophilic}} HOH_2N-C(CH_3)(CH_3)-O^{-}$$

rearranges

$$HO.N{=}C(CH_3)(CH_3) \xleftarrow[-H_2O]{\text{elimination}} HOHN-C(CH_3)(CH_3)-OH$$

The characteristic reaction of carbonyl compounds is *direct* nucleophilic addition, but mention should be made of addition reactions which are catalysed by acids. In these cases a proton will coordinate with the oxygen so that the electron deficiency at the carbon atom is now at a maximum and nucleophilic addition (as a second-stage reaction) is facilitated, thus:

$$\overset{\delta+}{C}{=}\overset{\delta-}{O} \xrightarrow[\text{H}^+]{\text{step (1)}} \overset{+}{C}{-}O{-}H \xrightarrow[\text{X}^-]{\substack{\text{step (2)} \\ \text{nucleophilic} \\ \text{addition}}} X{-}C{-}O{-}H$$

The degree of chemical reactivity shown by HCHO, CH_3CHO and CH_3COCH_3 decreases in the order given because the electron-releasing nature of the methyl groups reduces the positive character of the carbonyl carbon atom. The bulk of the methyl groups reduces ready access to this atom; this steric factor also contributes to reduced activity.

(d) REACTIONS OF THE BENZENE RING

Monosubstitution

Benzene contains a delocalized π-bond so that attack by electrophiles such as the nitronium ion, NO_2^+, does not result in the simple electrophilic addition described on page 244.

A mixture of concentrated nitric and sulphuric acids is used for the mononitration of benzene at temperatures below 55°C. Spectroscopic analysis confirms the presence of the nitronium ion (NO_2^+) in the nitrating mixture:

$$HNO_3 + 2H_2SO_4 \rightleftharpoons NO_2^+ + H_3O^+ + 2HSO_4^-$$

This electrophile will interact with the delocalized π-electrons in benzene producing a transitory complex (σ-complex). Proton loss rapidly occurs, so that a delocalized π-bond involving six electrons (aromatic sextet) is reformed and the associated stability of this arrangement is regained:

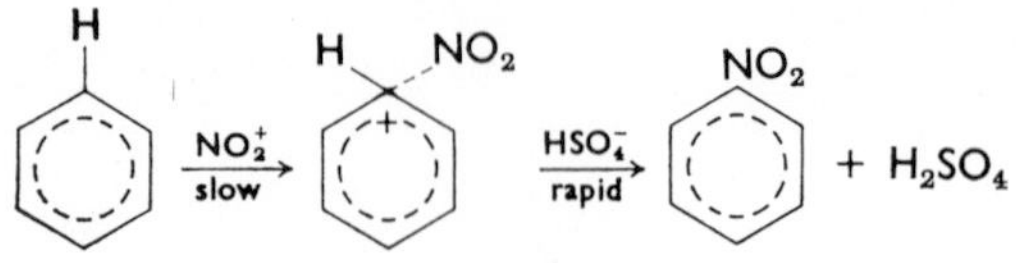

The function of sulphuric acid then, is to facilitate the production of nitronium ions. It is not surprising therefore that other strong acids can be used instead of sulphuric acid in the nitrating mixture.

The precise mechanism of sulphonation still gives rise to debate as to whether the electrophile is free sulphur trioxide, SO_3, or the bisulphonium ion $^+SO_3H$. There is currently more evidence supporting the view that the molecule produced thus:

$$2H_2SO_4 \rightleftharpoons SO_3 + HSO_4^- + H_3O^+$$

is the effective electrophile. The sulphur atom in SO_3 is electron deficient:

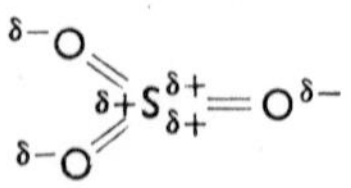

and will attack the electrons in the ring forming a transitory σ complex. Proton loss can occur producing a molecule containing the aromatic sextet.

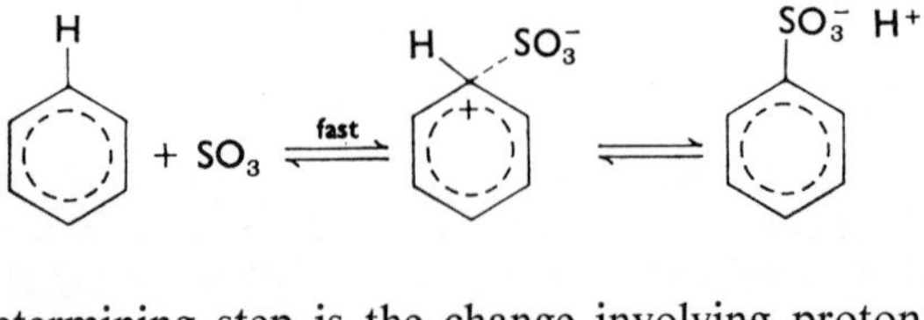

The rate-determining step is the change involving proton loss. This contrasts with nitration where the rate-determining step is attack by NO_2^+ ion. Again, the reactions are reversible and sulphonic acid in steam will produce benzene. Fuming sulphuric acid, which contains a high concentration of SO_3, is effective at room temperature and is generally preferred to hot concentrated sulphuric acid.

Disubstitution

When a mono-derivative of benzene is converted to a di-derivative three possible isomeric di-derivatives may be formed, thus:

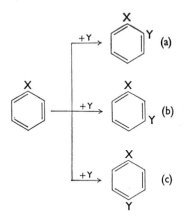

The di-derivatives (a), (b) and (c) are referred to as *ortho-*, *meta-* and *para-* di-derivatives respectively. Examples are:

meta-dinitrobenzene   *para*-chlorotoluene

The position taken by the second substituent depends upon the nature of the group already present in the ring. The directing influence of the first substituent group is well known. Experimental evidence indicates that the directing effect of the various groups is as indicated in Table 11.3.

Meta-*directing*	—NO₂	—SO₃H	—CHO	—COOH	—COOR	—CN
Ortho- *and* para-*directing*	—OH	—NH₂	—Cl	—Br	—CH₃	—C₆H₅

Table 11.3

Several attempts have been made to express the experimental evidence in terms of rules related to the nature of the directing group. The rules do not have any sound theoretical basis, nor do they attempt to explain in any detailed terms the mechanisms involved in the formation of a di-derivative. At the present time the electronic theory of valency gives the best interpretation of the directing influence involved in di-substitution.

Groups which are most readily substituted into a benzene ring are electrophiles (electron-seeking groups). They attack the molecule at positions of high electron density. If a group already present in a benzene ring causes particular positions in the ring to have increased electron density, attack in these positions is facilitated.

The effect of a substituent group on electron density at various positions in the ring depends upon combined mesomeric and inductive effects. Each effect will be discussed separately before the results of simultaneous action are indicated.

(i) *Mesomeric effect*

This can be conveniently discussed under two headings:

(*a*) *Groups containing unshared pairs of electrons* Examples are $-NH_2$, $-OH$ and $-Cl$, i.e. the groups found experimentally to be *ortho*- and *para*-directing. For such groups electron flow into the delocalized π-bond of the ring occurs, producing high electron density in the positions shown below:

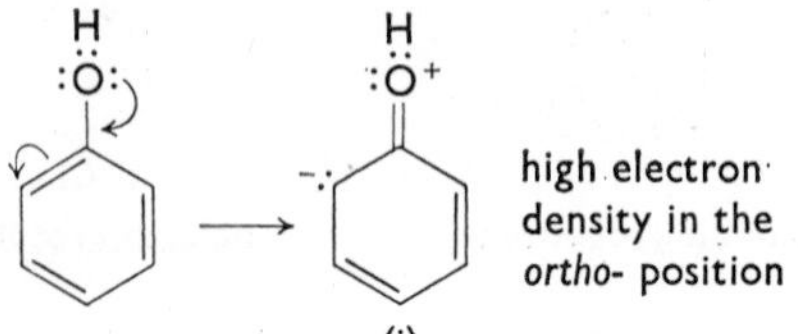

high electron density in the *ortho*- position

(i)

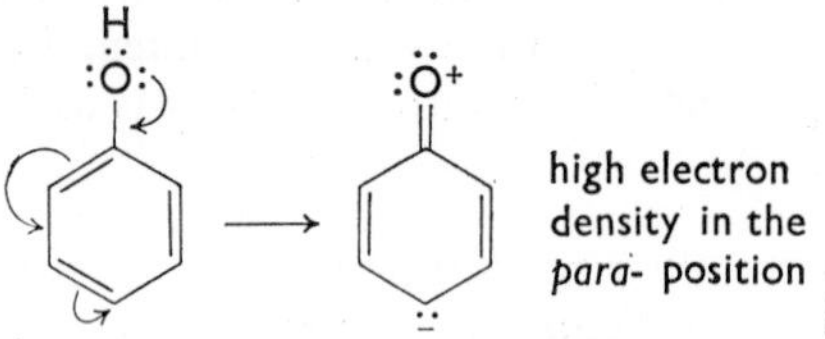

high electron density in the *para*- position

(ii)

The structure of phenol may be more accurately represented as a resonance hybrid of several structures, but structures (i) and (ii) indicate the possibility of high electron density in the *o*- and *p*- positions. The ring is said to be *activated* since attack by electrophilic reagents occurs more readily than with benzene itself.

(*b*) *Groups containing a double bond* Examples are $-NO_2$ and $-CHO$, i.e. the groups found experimentally to be *meta*-directing. In this case the mesomeric effect takes place in the reverse direction. Electrons flow out of the ring-producing positions of low electron density, thus:

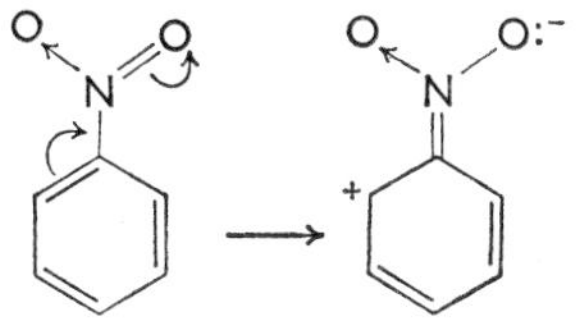

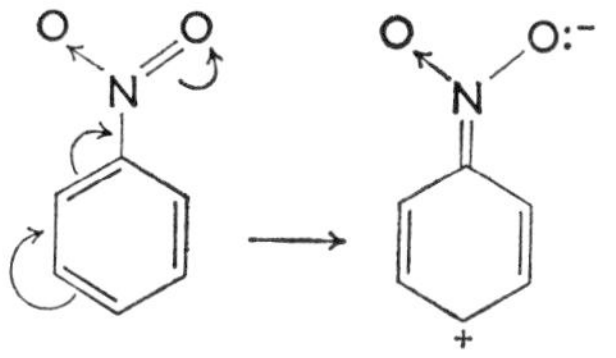

Nitrobenzene is a resonance hybrid of all possible structures, but (i) and (ii) show the possibility of low electron density in *o*- and *p*- positions and hence relatively high electron density in the *m*- position. The ring is *deactivated*; i.e. attack by electrophilic reagents occurs less readily than with benzene itself.

The attacking reagent itself can increase the chances of electron shift in the molecule by reinforcing the permanent mesomeric effect. This electromeric effect has been mentioned already on p. 228.

(ii) *Inductive effect*

The inductive effect of various groups has been discussed in detail on p. 227. Those groups which have a high affinity for electrons, e.g. $-NO_2$, will draw electrons from the ring, producing low electron densities in the *o*- and *p*- positions, thus:

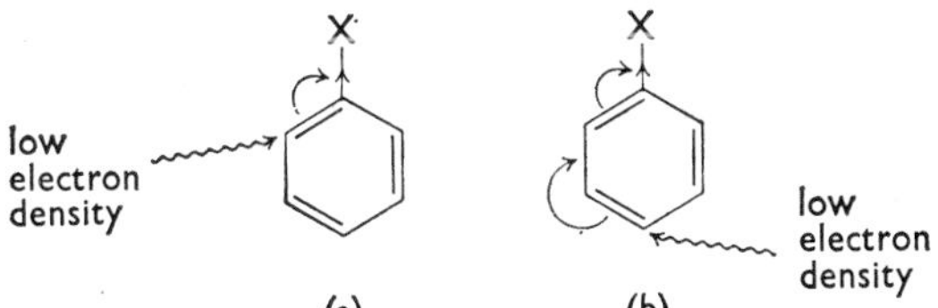

There is consequently a relatively high electron density in the *meta* position. Attack occurs in this position by electrophilic reagents, but the ring is deactivated; ease of attack being less than for benzene itself.

9+

Groups such as CH_3- and $(CH_3)_3 C-$ tend to lose electrons to the ring, producing high electron densities in the o- and p- positions, thus:

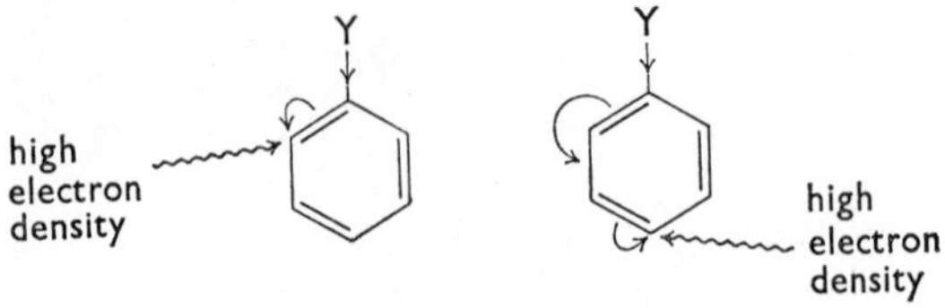

Attack by electrophilic reagent occurs in the o- and p- positions. The ring is activated so that substitution occurs more readily than with benzene.

Combination of mesomeric and inductive effects

Both effects acting simultaneously are responsible for the electron density distribution in a particular molecule. When the effects are not in opposition the directing influence and degree of activity (activated or deactivated molecule) is clear, as examples 1, 2, 3 and 6 in Table 11.4

Substance	Mesomeric effect of the substituent group	Inductive effect of the substituent group	Activity	Directing influence
1. $C_6H_5NO_2$	electron attracting	electron attracting	deactivated	m-
2. C_6H_5COOH	electron attracting	electron attracting	deactivated	m-
3. C_6H_5CHO	electron attracting	electron attracting	deactivated	m-
4. C_6H_5Cl	electron releasing	electron attracting*	deactivated	o- and p-
5. $C_6H_5NH_2$	electron releasing*	electron attracting	activated	o- and p-
6. $C_6H_5CH_3$	electron releasing	electron releasing	activated	o- and p-
7. C_6H_5OH	electron releasing*	electron attracting	activated	o- and p-

* Indicates the stronger effect.

Table 11.4

show. If the effects are in opposition the stronger of the two usually dictates *both* the orientating influence and the reactivity of the mono-derivative. This is seen to be true for examples 6 and 7 in the table. For chlorobenzene, however, the inductive effect is the stronger, giving a deactivated ring due to electron loss towards chlorine. This would normally produce low electron density in the *o*- and *p*- positions (as shown in diagrams (a) and (b), p. 249). Attack by an electrophilic (positive) reagent should therefore occur more readily at the *m*- position. The reagent itself, however, reinforces the mesomeric, *o*- and *p*- activating effect at the moment of attack (referred to as an electromeric effect; see p. 228). Therefore, whilst the activity of the ring is low, at the moment of attack whatever activity remains is concentrated at the *o*- and *p*- positions.

Mesomerism in toluene resulting in electron loss by the methyl group to the ring needs some explanation. Reference to the mesomeric effect discussed on p. 227 will show that groups can exhibit an electromeric effect if they possess unshared pairs of electrons or if they possess double bonds. Toluene does not fall into either category. The mesomeric effect in this case is thought to occur by a partial electron release from H to C in the C–H bonds of the methyl group (an effect referred to as *hyperconjugation*), thus:

(a)

(b)

The ring is activated and the methyl group is *o*- and *p*- directing ((a) and (b) are two of the several possible formal structures).

The above discussion suggests that a group already present in a benzene ring will show a tendency to direct subsequent groups into definite positions in the ring. It is important to note, however, that the formation of *all three di-derivatives* often occurs, though formation of the 'unfavoured' derivative(s) is reduced to zero or to a small percentage when the directing influence is strong, particularly if the ring is activated. The experimental evidence shown in Table 11.5 illustrates this point and gives ample support to the theory which has been discussed.

Reaction	Reactant	Activated or deactivated ring	Ease of reaction	YIELD		
				Ortho %	Para %	Meta %
Nitration	Mononitro-benzene	D	More difficult than benzene, fuming HNO_3 and conc. H_2SO_4	6	1	93
	Chlorobenzene	D	Difficult	30	70	—
	Benzoic acid	D	About the same as benzene	19	1	80
	Phenol	A	Readily, room temp. dilute HNO_3	40	60	—
	Toluene	A	Low temp. conc. HNO_3, readily	59	37	4
Sulphonation	Chlorobenzene	D	More difficult than benzene	low	high	—
	Benzoic acid	D	About the same as benzene	low	low	high
	Phenol	A	Readily	high	low	—
	Aniline	A	Readily	low	high	—
	Toluene	A	Readily	13	8	79
Bromination	Phenol	A	Readily	2:4:6 tri-derivative i.e. very high o- and p- influence		
	Aniline	A	Readily			
Chlorination	Toluene	A	Readily—room temp. catalyst	high	high	—
	Phenol	A	Readily—room temp.	2:4:6 tri-derivative		
	Benzoic acid	D	About the same as benzene	low	low	high

Table 11.5   Substitution reactions of monoderivatives

References

Books

A Guidebook to Mechanism in Organic Chemistry, P. Sykes (Longmans)
Basic Organic Chemistry, A Mechanistic Approach, J. J. Tedder and A. Neehvatal (Wiley)
An Introduction to Modern Chemistry, M. J. S. Dewar (Athlone Press)

Acids, Bases and the Chemistry of the Covalent Bond, C. A. Vanderwerf (*Selected Topics in Modern Chemistry* series) (Chapman and Hall)
Organic Reactions, G. Illuminate (in *Chemistry Today*, *A Guide for Teachers*, OECD Publication)
An Introduction to Electronic Theory of Organic Compounds, H. L. Heys (Harrap)
An Introduction to Electronic Theories of Organic Chemistry, G. I. Brown (Longmans)

Films and Filmstrips

Mechanism of an Organic Reaction, H. Rapoport (film), (available from Sound-Services Ltd., cat. no. 4166/999)

12. Polymers

Nature of polymers

Polymers consist of giant molecules produced by reaction between small molecules (monomers). Although the polymer may be an extremely large complicated molecule with a molecular weight of the order of 10^2 to 10^5, it is held together by simple covalent bonds. There are a great many naturally occurring polymers, such as rubber, starch, cellulose and proteins, which are produced by animal and vegetable metabolic processes. In the last thirty or forty years very many synthetic polymers have also been made. The synthesis and investigation of substances such as nylon, polystyrene, and polyvinyl chloride has shown that artificial and natural polymers are essentially similar in structure.

Most chemical reactions involve combination of two or three simple molecules to produce a molecule with a low molecular weight. Formation of a giant molecule depends on the functionality (that is the number of reactive centres)* of the monomer. When reacting species are only monofunctional a low-weight molecule is obtained; the formation of an ester is an example:

$$\underset{\substack{\text{aliphatic}\\\text{carboxylic}\\\text{acid}}}{R.COOH} + \underset{\substack{\text{aliphatic}\\\text{primary}\\\text{alcohol}}}{R'OH} \rightleftharpoons \underset{\text{ester}}{R.COOR'} + H_2O$$

Suppose, however, that each of the reacting monomers is difunctional; then the formation of a high-molecular weight linear polymer is possible, thus:

propylene   polypropylene

* Functionality often equals the number of functional groups; e.g. glycol $(CH_2 COOH)_2$ is difunctional. This is not always the case; ethylene, $CH_2=CH_2$, for example, is difunctional although it possesses only one functional group.

Chain length

Besides the formation of long-chain linear polymers there is also a possibility of rings being formed, but this is very low. Initially the number of chain ends present is very small compared to the number of monomers. The probability, therefore, of chain ends reacting with monomers is high. When very few monomers remain the reaction of two of the many separate chains is much more probable than the reaction of chain ends of the same molecule with one another. The chains formed will be of varying length; this point will be taken up later. A mathematical expression for degree of polymerization can be derived in the following way:

Suppose the number of molecules present initially is n_0, and the number of molecules present after t seconds is n_t. Each bond formed results in the loss of one molecule, and therefore

$$\text{total number of bonds formed after } t \text{ seconds} = (n_0 - n_t)$$

Two reacting centres are used to form a bond; hence

$$\text{total number of reacting centres used after } t \text{ seconds} = 2(n_0 - n_t)$$

The extent of reaction, E, may be defined thus:

$$E = \frac{\text{number of reacting centres used in } t \text{ seconds}}{\text{number of reacting centres present initially}}$$

Therefore

$$E = \frac{2(n_0 - n_t)}{n_0 f}$$

(where f is the functionality or number of reacting centres per molecule). Therefore,

$$E = 2\left(1 - \frac{n_t}{n_0 f}\right) \tag{1}$$

The degree of polymerization DP may be defined thus:

$$\text{DP} = \frac{\text{number of molecules present initially}}{\text{number of molecules present after } t \text{ seconds}}$$

Therefore,

$$\text{DP} = \frac{n_0}{n_t}$$

Substituting into (1) above,

$$E = 2\left(1 - \frac{1}{\text{DP} f}\right)$$

Hence the degree of polymerization is given by

$$DP = \frac{1}{(1 - \frac{1}{2}Ef)}$$

This expression for degree of polymerization can be used to illustrate one of the most important aspects of polymer chemistry. Suppose the yield of product is 98 per cent in a polymerization process involving bifunctional molecules. The values of E and f are 0·98 and 2 respectively, giving a DP value of 50. Values of DP in polymerization processes, if they are to be effective, should be above 500. This means that percentage purity of the reacting species should be very high indeed.

Molecular weights

It was mentioned earlier that the lengths of chains formed varied. The values for the weights of various molecules present in a polymer will show a Maxwellian distribution (Fig. 12.1). Measurement of molecular

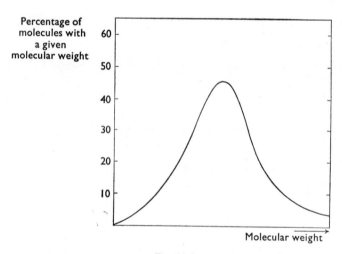

Fig. 12.1

weight by some methods will give an average weight value. Values for spread of molecular weight in a given sample are important and can be calculated from experimental data.

Branched and cross-linked polymers

Only linear polymerization has so far been discussed. This was seen to occur when the monomer molecules were difunctional. When trifunctional monomers are used, branched and cross-linked polymers can be obtained. Suppose, for example, phenol and formaldehyde are the reacting monomers, then linear polymerization occurs with the elimination of molecules of water:

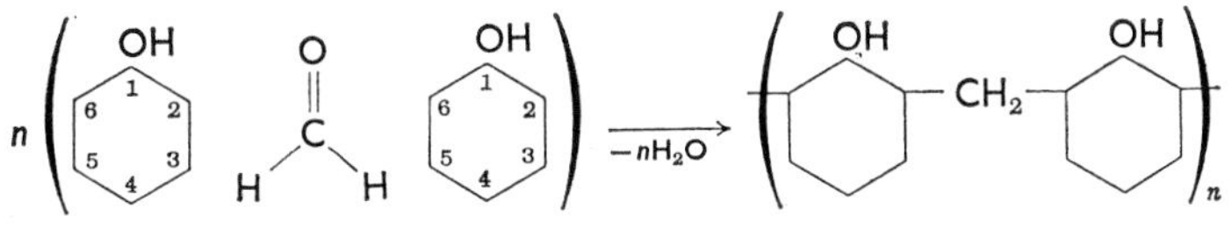

Hydrogen atoms in positions 2 and 6 are eliminated. But by elimination of hydrogen atoms in position 4 as well, branching and cross-linking occur, giving the structure shown in Fig. 12.2.

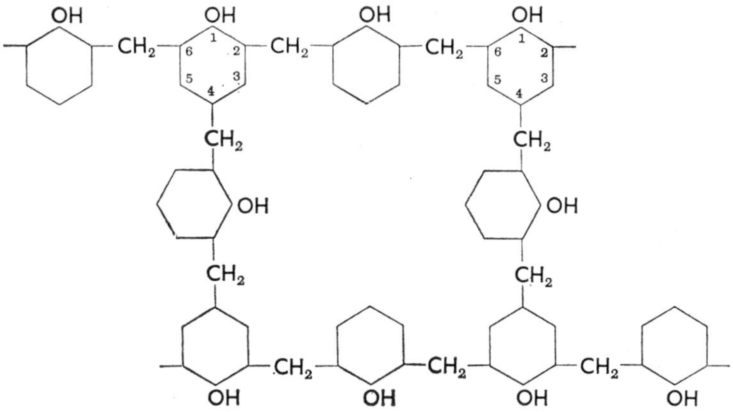

Fig. 12.2

Note: Here, and on subsequent pages, double bonds have sometimes been omitted in aromatic compounds to simplify the diagrams.

Types of polymerization

Addition polymerization

Nearly forty years ago the first vinyl polymer was made; since then many others have been obtained. (Strictly speaking the term *vinyl* is used to refer to those compounds containing the group $CH_2=CH-$. However, in the field of polymer chemistry the term is commonly used to describe monomers of the type $CR_2=CR_2$ and polymers formed from them.)

Using relatively simple techniques, long-chain polymers are obtained by self-addition of vinyl monomers; the process is called *addition*

9*

polymerization. Table 12.1 below shows monomers and the corresponding polymers which have been obtained from them.

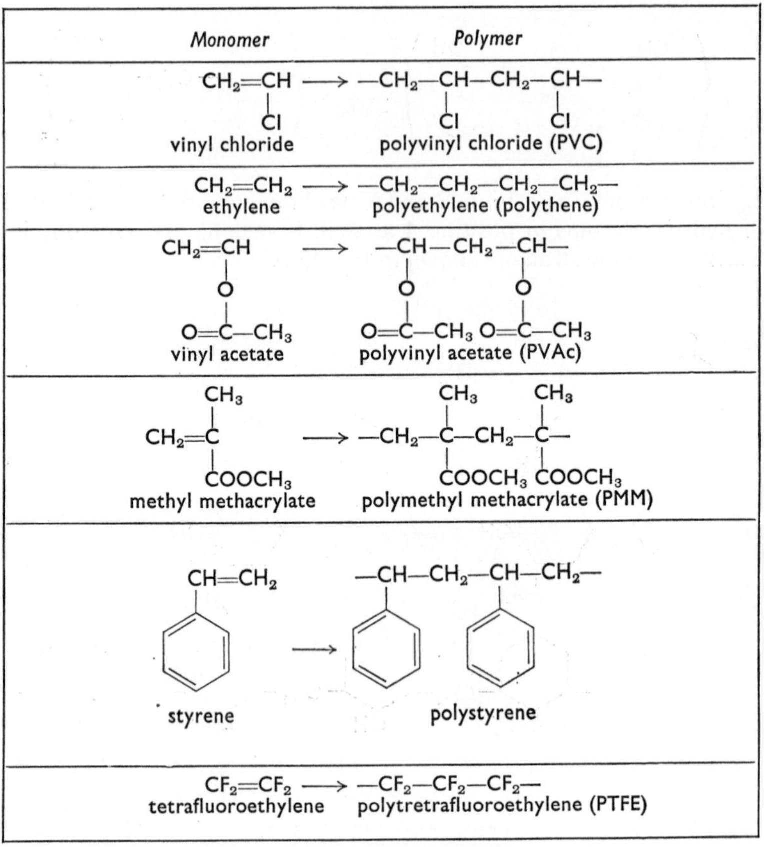

Monomer	Polymer

CH_2=CH
|
Cl
vinyl chloride
	—CH_2—CH—CH_2—CH—
	   | |
	   Cl Cl
	polyvinyl chloride (PVC)

CH_2=CH_2 ethylene	—CH_2—CH_2—CH_2—CH_2— polyethylene (polythene)
CH_2=CH | O | O=C—CH_3 vinyl acetate	—CH—CH_2—CH— | | O O | | O=C—CH_3 O=C—CH_3 polyvinyl acetate (PVAc)

CH_3
|
CH_2=C
|
$COOCH_3$
methyl methacrylate
	CH_3 CH_3
	| |
	—CH_2—C—CH_2—C—
	     | |
	  $COOCH_3$ $COOCH_3$
	polymethyl methacrylate (PMM)

CH=CH_2 ⬡ styrene	—CH—CH_2—CH—CH_2— | | ⬡ ⬡ polystyrene
CF_2=CF_2 tetrafluoroethylene	—CF_2—CF_2—CF_2— polytetrafluoroethylene (PTFE)

Table 12.1

Techniques used in making vinyl polymers
Most of the vinyl polymers mentioned are made using one of the three techniques described below:

(*a*) *Mass polymerization*   The liquid monomer is heated and stirred, small amounts of a substance which promotes polymerization (a catalyst) being added. The liquid thickens and eventually solidifies on cooling. Careful control is necessary to avoid premature solidification.

(*b*) *Solution polymerization*   The compound is dissolved in a solvent. A catalyst is added to the mixture which is heated and stirred. This is a good method since the polymer remains in solution and control is much easier.

(c) *Emulsion polymerization* Here the monomer is stirred with water and an emulsifying agent. Provided that a catalyst is used, polymerization is rapid even at low temperature. When the emulsifying agent is neutralized chemically, the polymer particles coagulate.

(d) *Pressure processes* Some polymers, e.g. polyethylene, are produced by application of high pressures. In other polymerizations, e.g. production of PVC, pressure is used together with one of the techniques such as emulsification already mentioned.

The state of the compound during the polymerization process passes through the molecular changes illustrated in Fig. 12.3.

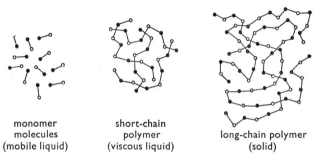

| monomer molecules (mobile liquid) | short-chain polymer (viscous liquid) | long-chain polymer (solid) |

Fig. 12.3

Vinyl polymers are usually thermoplastics. This means that they can be softened and reshaped by heating. When softened they flow easily and mould well. They are light, often transparent or translucent, tough and resilient. Some vinyl polymers are hard and brittle, but by using an additive, called a plasticizer, they can be transformed into pliable material. The properties and uses of the better-known vinyl polymers are summarized in Table 12.2.

The reader's attention is drawn to the very high resistance to heat and solvent action shown by PTFE. This substance will withstand temperatures of over 300°C indefinitely. It is unaffected by the prolonged action of either hot concentrated sulphuric acid or aqua regia. The chemical and thermal stability of this plastic is due to the presence of fluorine. When this extremely reactive element is combined with certain other elements, compounds showing remarkable inertness are produced.

	Polythene	Polyvinyl acetate (PVAc)	Polyvinyl chloride (PVC)	Polymethyl methacrylate (PMM)	Polytetra-fluoro-ethylene (PTFE)	Polystyrene
Appearance	White, translucent	Clear, colourless	Varies	Clear, transparent	White, translucent	Clear, colourless
General physical properties	Odourless, tasteless, tough, flexible	Odourless, tasteless, adhesive (tacky)	Unplasticized, brittle, plasticized, flexible	Odourless, tasteless, tough, rigid	Tough, flexible, waxy feel, self-lubricating	Tasteless, odourless, rigid, brittle
Effect of heat	Thermo-plastic. Softens in boiling water. Burns slowly	Thermo-plastic	Thermo-plastic. Softens in boiling water. Does not burn well	Thermo-plastic. Softens in boiling water. Burns slowly	Good resistance to high temp. Does not burn	Thermo-plastic. Softens in boiling water. Burns slowly
Solvents, acids and alkalis	Unaffected by house-hold substances. Good resistance to solvents, acids, alkalis	Soluble in many organic solvents, e.g. ace-tone and alcohol	Resists attack by solvents, acids and alkalis	Unaffected by household substances, weak acids and alkalis	No known solvent. Not attacked by acids or alkalis	Unaffected by household substances except citrus fruits. Attacked by petrol and turps. Soluble in benzene
Light	Deterior-ates with long exposure	Little effect	Colour deteriorates	Little effect	No effect	High trans-parency. Does not deteriorate
Electrical	Excellent insulator	—	Good insulator	Excellent insulator	Excellent insulator	Excellent insulator
Uses	Sheet—for packaging and insulating. Tube—for plumbing. Powder—for moulding	Solutions—as adhesives, laquers and paints. Powder—for moulding	Sheet—for packaging and rainwear. Tube—for piping. Powder—for moulding	Sheet—for ornamental glazing. Tube—for electric lighting. Powder—for moulding, e.g. dentures	Used as film, sheet, tube and rod. For resistant coating, liners and bearings	Solution—for painting. Film and sheet for insulating. Powder—for moulding. Foam—expanded polystrene as insulator

Table 12.2

Condensation polymerization

(Phenyl-formaldehyde polymers, polyamides and polyesters)
Condensation polymerization is a process in which simple molecules combine to give a giant molecule containing a large number of repeating units. During the process simple molecules such as water are eliminated.

About 1907 Baekeland heated together phenol and a solution of formaldehyde (formalin) with a small trace of alkali. He obtained a thick viscous liquid which solidified when cooled. The solid obtained was called Bakelite. It was hard and brittle and could be ground into a powder. In this state it is a thermoplastic and can be moulded, but unlike the vinyl polymers continued heating causes it to set (setting with the action of heat is called *thermosetting*). The difference between the two solids obtained is due to a difference in structure. The initial action of heat is to produce long-chain polymers, thus:

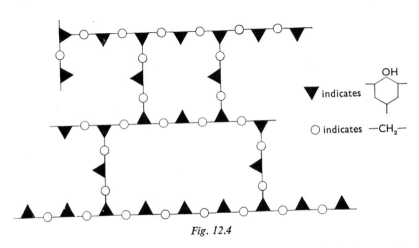

As explained on p. 257, molecules of phenol and formaldehyde have combined, with the elimination of water. When cool the tangled mass of chains becomes a solid. Further heating, however, causes side chains and cross-links to form; this was explained on p. 257 and is also indicated in Fig. 12.4. Eventually so much cross-linking occurs that the polymer becomes solid.

Fig. 12.4

Manufacturing processes for polymers of this kind include the use of additives called *fillers*. Using these it is possible to modify the properties of the polymer and extend its usefulness. Common fillers used are asbestos, carbon, graphite and paper. The polymers incorporating these substances have good heat resistance, good electrical conductivity, good resistance to wear and good resistance to impact, respectively.

Condensation polymers, giving plastics with a wide variety of uses, are obtained from formaldehyde and each of the cresols shown below:

ortho-   meta-   para-

cresols

They are also obtained from formaldehyde and melamine, which has the structure shown:

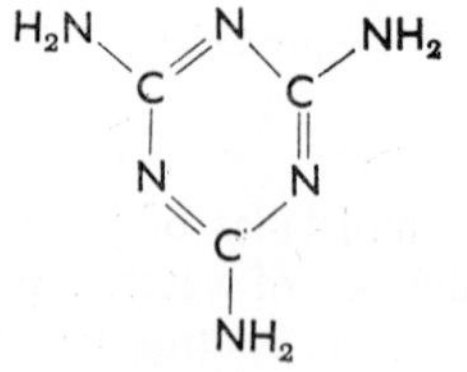

melamine

Nylon

This term is used to describe polyamides. These are polymers produced by condensation polymerization, as was Bakelite. Various kinds of nylon polyamides can be obtained by condensing a diamine with a dibasic acid. Thus, if hexamethylene diamine and adipic acid are mixed, hexamethylene diammonium adipate (called nylon salt) is obtained. A solution of this salt is heated, allowing water to distil away. The temperature is raised to about 300°C, when the nylon is obtained as a molten mass. It is forced through a small orifice and cooled to give solid thread.

$$2n\{NH_2(CH_2)_6NH_2 + HOOC(CH_2)_4COOH\}$$

hexamethylenediamine   adipic acid

$$\xrightarrow[-4nH_2O]{} (-NH(CH_2)_6NHCO(CH_2)_4CONH(CH_2)_6NHCO(CH_2)_4CO-)_n$$

polyhexamethylene adipamide (nylon 6:6)

Nylon made from these compounds is called nylon 6:6 since the monomers each contain six carbon atoms. By using sebacic acid (with ten carbon atoms) instead of adipic, nylon 6:10 is obtained thus:

$$2n\{NH_2(CH_2)_6NH_2 + HOOC(CH_2)_8COOH\}$$

hexamethylene diamine   sebacic acid

$$\xrightarrow[-4nH_2O]{} (-NH(CH_2)_6NHCO(CH_2)_8CONH(CH_2)_6NHCO(CH_2)_8CO-)_n$$

polyhexamethylene sebacamide (nylon 6:10)

The properties and uses of the different kinds of nylon are well known.

Terylene (Dacron)

Polymerization condensation using a dihydric alcohol and a dibasic acid (in a similar manner to the production of nylon) gives, by elimination of water, a long-chain polymer containing a repeating ester group

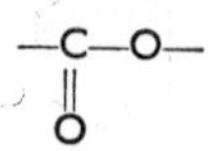

$$-\underset{\underset{O}{\|}}{C}-O-$$

The polyester is a liquid, but by combination with styrene, which also polymerizes, cross-linking is achieved and a solid polyester is obtained.

Polyesters of this kind are used for making fibreglass car bodies and boat hulls. Various kinds of polyesters have been made; one of the newer polyester fibres is spun to give the thread known as Terylene.

Mechanism of polymerization

Free-radical polymerization

There is strong evidence that some vinyl polymers are produced by a free-radical mechanism. In the initiation step of the chain reaction the additive (benzoyl peroxide is frequently used), at a certain temperature, will produce free radicals thus:

Initiation
$$(C_6H_5COO)_2 \longrightarrow 2C_6H_5COO\cdot$$
$$C_6H_5COO\cdot \longrightarrow C_6H_5\cdot + CO_2$$

Propagation steps involve the addition of a radical $(R\cdot)$ to the monomer, followed by further addition of monomer to the new radical, thus:

Propagation
$$\underset{\text{vinyl compound}}{CX_2{=}CH_2} + R\cdot \longrightarrow \underset{\text{new radical}}{\cdot CX_2{-}CH_2{-}R}$$
$$CX_2{=}CH_2 + \cdot CX_2{-}CH_2{-}R \longrightarrow \cdot CX_2{-}CH_2{-}CX_2{-}CH_2{-}R$$

Termination occurs when two chain radicals combine, or when a chain radical combines with an initiator radical thus:

Termination $R\rightsquigarrow CH_2CX\cdot + \cdot CXCH_2\rightsquigarrow R \longrightarrow R\rightsquigarrow CH_2CXCXCH_2\rightsquigarrow R$

or $R\rightsquigarrow CH_2CX\cdot + \cdot R \longrightarrow R\rightsquigarrow CH_2CXR$

Ionic polymerization

(i) *Cationic mechanism*

This occurs in the polymerization of vinyl compounds such as styrene and isobutylene. A carbonium ion, $\underset{/}{\overset{\backslash}{C^+}}$, is produced in the initiation step because of the presence of strong Lewis acids such as $AlCl_3$, $AlBr_3$ and BF_3; this may be represented as follows:

Initiation $AlCl_3 + H_2O + \underset{\text{isobutylene}}{CH_2{=}C(CH_3)_2} \longrightarrow CH_3{-}\overset{+}{C}(CH_3)_2 + [AlCl_3(OH)]^-$

Propagation consists in the continued addition of monomer molecules to the carbonium ion, as shown below:

Propagation $CH_3{-}\overset{+}{C}(CH_3)_2 + CH_2{=}C(CH_3)_2 \longrightarrow$

$$CH_3{-}C(CH_3)_2{-}CH_2{-}\overset{+}{C}(CH_3)_2$$

Termination occurs when the long-chain polymer loses a proton, thus:

Termination $CH_3\text{---}CH_2\text{---}\overset{+}{C}(CH_3)_2 + [AlCl_3(OH)]^- \longrightarrow$

$$CH_3\text{---}CH_2\text{---}\overset{\overset{\displaystyle CH_2}{\|}}{C}\text{---}CH_3 + [AlCl_3(OH)]^-H^+$$

It should be noted that there is only a small trace of water present (the water here is what is known as a *co-catalyst*). Reaction is usually carried out in an organic solvent, so that ionization is slight and carbonium ion and cation remain in close proximity throughout polymerization.

(ii) *Anionic mechanism*

The initiator in this kind of mechanism is a Lewis base. Compounds used include sodium and potassium amides and hydrides in a solvent such as liquid ammonia. The initiation step involves combination of the Lewis base with the monomer to give a carbanion $\overset{}{\underset{}{\diagup}}C^-$. In propagation, successive additions of monomer molecules to the carbanion produce a long-chain polymer. Termination involves proton gain by the long-chain polymer. These changes may be represented thus:

Initiation $\qquad\qquad MA \rightleftharpoons M^+ + A^-$

$$CR_2{=}CH_2 + A^- \longrightarrow \bar{C}R_2\text{---}CH_2\text{---}A$$
$$\underset{\text{carbanion}}{}$$

Propagation $CR_2{=}CH_2 + \bar{C}R_2\text{---}CH_2\text{---}A \longrightarrow \bar{C}R_2\text{---}CH_2\text{---}CR_2\text{---}CH_2\text{---}A$

Termination
$^-CR_2(\text{---}CH_2\text{---}CR_2\text{---})CH_2\text{---}A + NH_3 \longrightarrow CHR_2(\text{---}CH_2\text{---}CR_2\text{---})_nCH_2\text{---}A + NH_2^-$
$\qquad\qquad\qquad\underset{\text{solvent}}{}$

In ionic polymerization processes, cationic reagents have been used more extensively, but the number of anionic polymerizations being investigated is beginning to increase.

Conclusion

The material presented in this chapter is intended only as an introduction to polymer chemistry. We have not touched upon certain subjects, such as the naturally occurring polymers. Several inorganic polymers and polymeric ions, however, have been discussed in the chapters on the general chemistry of the elements.

References

Books

The Nature and Chemistry of High Polymers, K. O'Driscoll (Chapman and Hall)

Polymers and Polymerisation, V. Crescenzi (in *Chemistry Today, A Guide for Teachers*, OECD publication)

Your Guide to Plastics, J. G. Cook (*Science for Everyman Series*, Merrow Publishing Co.)

Textbook of Organic Chemistry, L. N. Ferguson (Van Nostrand)

Index

AB close packing 72
ABC close packing 72
Acetic acid, structure of 226
Acetylene molecule 45
Activation analysis 97
Addition 234
Addition polymerization 257
Addition reactions 243
Alcohols 232
Alkali metals 105–9
  mobility of ions of 107
  peroxides of 107
Alkaline earth metals 109–16
Alkanes 235
Alkyl halides 240
Allowed energy states 8, 11
Alpha-particles 3
  scattering of 5
Alpha-rays 2
'Aluminate' ions 121
Aluminium chloride 119
Amines 232–3
Ammonia molecule 36, 39, 136
Asymmetry 206–7
Atomic number 6, 7
Atomic theory 1
'Aufbau' principle 19
Average bond energy 32

Band model of solids 84–5
Benzene, disubstitution products of 247
  monosubstitution of 245
  structure of 47–9, 226
Beryllium, anomalous properties of 111
  and aluminium 114
Beryllium dichloride 44–5

Beta-rays 2
Binding energy 92
Black phosphorus 143
Body-centred cubic packing 73
Bohr atom 8
Bond stability, conditions for 32
Borazole 122
Born-Haber cycle 29
Boron, anomalous properties of 121
  hydrides of 122
  and silicon 125
Boron Group (III) 117–25
Boron nitride 123
Boron trifluoride 40, 119

Cadmium iodide structure 198
Caesium chloride, structure of 80
  unit cell of 77
Carbon, anomalous properties of 131
  and silicon 132
Carbon dioxide 46–7, 128
Carbon Group (IV) 125–35
Carbon monoxide 127
Carbonium ion 236
Carbonyl compounds 244
Carboxylic acids 230
Catalytic activity 180
Catenane 215
Cathode rays 4
Chelates 219
Chemical topology 214
Chlorine 33
Chlorine isotopes 5, 7
Chromium 190
Clathrate compounds 171
Close packing of spheres

Cobalt   193
Coinage metals   195–7
Colour in transition element compounds
   186–9
Complexes   180–9, 217–23
Conformational analysis   212
Conformational isomers   212
Co-ordinate bonding   32, 49
Co-ordination compounds   217
Co-ordination isomerism   193
Co-ordination number   71, 74, 182
Copper   195–7
   electronic structure of   61
Covalent bonding   31, 32
Covalent crystals   82
   properties of   83
Crystal habit   70
Crystal lattice   70
Cubic close packing   72
Cyclohexane, boat form   213
   chair form   213
Cyclohexane dicarboxylic‾acids   210
Cyclopropane dicarboxylic acids   209

Dative covalency   32, 49
d-block elements   177–89
de Broglie equation   10
Delocalization   47, 48
Delocalized π-bond   47, 48, 123, 226
Dewar   225
Diagonal relationships   108, 114, 125
Diamond, properties of   132
Diamond structure   82
Diastereoisomers   210
Dichloroethylene   204
Directing influence of substituent groups
   247
Dissociation, heat of   29
Disubstitution of benzene   247

E1, E2 reactions   239
Einstein's equation   10
Electromeric effect   228
Electron   3, 90
Electron affinity   29, 61
   table of values   62
Electron field, structure of   8
Electron pair repulsion theory   172

Electronegativity   62
   table of values   63
Electrophile   224
Electrophilic addition   243
Electrovalency   27
Elimination   234
Elimination reactions   239
Enantiomorphism   206
Endoergic change   93
Ethylene molecule   43
Exoergic change   93
Extravalency orbitals   32

Fajan's rules   69
Fischer projection   207
Fluorine, anomalous properties of
   167
Fluorite structure   81
Free radicals   235
Frenkel defect   86
Fumaric acid   204

Gamma-rays   2
Geometrical isomerism   193, 203
Graphite, properties of   130
   structure of   83
Group   22

Halides, of Group V   137, 143
   of Group VI   156–7
Halogen oxides   164
Halogen oxy-acids   165
Halogenoids   168
Halogens, oxidizing power of   159–60
   physical properties of   158
   reactions with alkalis   161
   reactions with water   160
Helium, liquid   170
Heterolysis   235
Hexacyanoferrate (III) ion   181, 184
Hexagonal close packing   72
Homolysis   235
Hybridization   36
Hydrate isomerism   195, 222
Hydrides, of boron   121
   of Group IV   127
   of Group V   136–7

Hydrides, of Group VI 150–1
  of Group VII 163
Hydrogen, isotopes of 105
  position in Periodic Table 104
Hydrogen bonding 51, 103, 149, 163
  in hydrogen fluoride 103
  in water 52
Hydrogen chloride molecule 34
Hydrogen ion 102
Hydrogen molecule 33
Hydrogen molecule ion 31
Hydrogen peroxide 153
Hydroxonium ion 103

Ice structure 151–2
Inductive effect 227, 231, 249
Inert pair effect 118
Infra-red spectroscopy 211
Interhalogen compounds 166–7
Interstitial sites 74
Inversion 239
Iodine cations 159
Iodine crystal 84
Iodine heptafluoride 162
Ionic bond 27
Ionic compounds, properties of 82
Ionic crystals 79–82
Ionization, heat of 29
Ionization potential (energy) 55–6
  factors affecting 56–7
  periodic variation of 58–61
Iron 192
Iron (II) oxide 87
Isomerism 202
  co-ordination 193, 223
  hydrate 193, 222
  ionization 222
Isomers 202
  conformational 212
  geometrical 193, 203
  optical 193, 206
  stereo- 203–19
  structural 202
  topological 214
Isotope dilution 99
Isotopes 5, 7

Kekulé 225

Lactic acid 206
Lanthanides 23
Lattice defects 86–8
Lattice energy 29
Laws of chemical combination 1
Ligand 180, 219
Ligand field theory 182–6
Line spectrum 8
Lithium, atypical properties of 108
  and magnesium 108

Magnesium, zinc and beryllium 114
Maleic acid 204
Maleic anhydride 204–5
Manganese 191
Mass number 7
Mass spectrograph 5
Maximum multiplicity 25
Mendeleev 17
Mercury (I) ion 199–200
Mesomeric effect 227, 248
Mesomerism 225
Metaborates 124
Metallic bonding 51
Metallic crystals 78–9
Metals, properties of 79
Methane molecule 38
Molecular crystals 83–4
Molecular shape 35
Monosubstitution of benzene 245
Multiple bonds 43

Neutron 3, 90
Newlands, John 17
Nickel 194
Nickel carbonyl 128
Nickel (II) oxide 88
Nicol prism 206
Nitrate ion 140
Nitration of benzene 245
Nitric acid 139, 140
Nitrite ion 140
Nitrogen, anomalous properties of 140–4
  electronic structure of 60
  and phosphorus 144
Nitrogen Group (V) 135–45
Nitrogen oxides 138

Nitrogen trichloride 143
Nitrous acid 139
Noble gas compounds 170–3
Noble gases 169–73
Non-stoichiometric defects 88
Non-stoichiometry 88, 180
Nuclear equations 8
Nuclear magnetic resonance 211
Nuclear structure, models of 93–5
  liquid drop model of 93
  shell model of 93
Nucleon 90
Nucleophile 224
Nucleophilic addition 244
Nucleus 90–100
  diameter of 91
  structure of 6, 93–5
Nylon 262

Octahedral ligand field 182–4
Octahedral site 74
Open packing of spheres 71
Optical isomerism 193, 205
Orbital overlap 31
Orbitals 15–16
Oxygen Group (VI) 145–57
Oxygen molecule 145
Ozone molecule 146

Packing of spheres 71–5
Paramagnetism 186
Pauli Exclusion Principle 17
Pauling 63
p-block elements 117–74
Period 21
Periodic Classification 17, 22
Periodic Law 17
Phenol 231
Phosphorus oxides 139
Phosphorus oxy-acids 141
Phosphorus pentachloride 137
Phosphorus trichloride 143
Pi-bonding 44
Pi-complex 229
Planck's constant 8
Polarized light 205
Polyethylene 258
Polyhalide ions 162

Polymers 254
Polymerization, mechanism of 263
  methods of 258
  types of 257
Polypropylene 254
Polytetrafluoroethylene (PTFE) 259
Polyvinyl chloride 258
Positive ray analysis 4
Positive rays 4
Positron 90
Probability distribution 12–14
Proton 3, 90
Pseudo-halogens 168

Quantum 8
Quantum numbers 9, 10, 18
Quartz 129

Racemization 210
Radioactive fission 99
Radioactive fusion 99
Radioactivity 95–6
  induced 3
  natural 2
Radio-carbon dating 99
Radio-isotopes 96
Radius, covalent 66
  ionic 64
  metallic 64
  Van der Waals 66
Radius ratio 75
Reaction mechanism 234
Rearrangement 234
Red phosphorus 143
Resolution 210
Resonance 225
Rutherford 5

s-block elements 101–16
Scandium 189
Schottky defects 86
Semi-conductors 85, 88
Sigma-bonding 44
Sigma complex 245–6
Silicates 133
Silicon dioxide 129
Silicones 134

S_N1 reactions 236
S_N2 reactions 236
Sodium chloride structure 28, 79
Sodium chloride unit cell 76
Sodium hydride 102
Solid state 70–89
sp hybridization 41
sp^2 hybridization 39
sp^3 hybridization 36
sp^3d hybridization 42
sp^3d^2 hybridization 42.
Stereoisomerism 203
Stoichiometric defects 86
Structural isomers 202
Sublimation, heat of 29
Substitution 234
Substitution reactions 235
Sulphate ion 154
Sulphonation of benzene 246
Sulphur dioxide 153
Sulphur hexafluoride 148
Sulphur molecule 146–7
Sulphur oxy-acids 155
Sulphur trioxide 154

Tartaric acids 206–8
Tellurium tetrachloride 148
Terylene 262
Tetrahedral ligand field 185
Tetrahedral site 74
Thiosulphate ion 154
Titanium 190

Topology 214
Transition elements 23, 175
    melting points of 178
Trigonal site 74

Unit cell 75
Unit charge 3

Vanadium 190
Van der Waals bonds 53
Variable oxidation state 178–80
Vinyl chloride 258
Vinyl polymers 257

Walden inversion 239
Water molecule 35, 149
Wave mechanics 12
Werner 217
White phosphorus 143
Wurtzite structure 81

Xenon, compounds of 172–3

Zeeman effect 10
'Zincate' ions 108
Zinc blende structure 80
Zinc, electronic structure of 61
Zinc Group 197–200
Zinc oxide 88